普通高等教育"十一五"国家级规划教材

NEW ESSENTIAL
COLLEGE ENGLISH

新起点

〔修订版〕

新起点
大学基础英语教程

总主编：杨治中

主　编：王海啸

副主编：韩　旭

编　者：（以姓氏笔画为序）

王海啸	刘　芳	孙爱娣
吴淑云	张丽萍	张沂昀
张晓红	邵继荣	倪辉莉
黄　硕	黄　燕	韩　旭
傅　俭		

4

读写教程

外语教学与研究出版社
FOREIGN LANGUAGE TEACHING AND RESEARCH PRESS
北京　BEIJING

图书在版编目(CIP)数据

新起点大学基础英语教程. 读写教程. 4 / 杨治中主编；王海啸分册主编；王海啸等编 .—
修订本 .— 北京：外语教学与研究出版社，2009.8
ISBN 978 - 7 - 5600 - 8948 - 5

Ⅰ. 新…　Ⅱ. ①杨…②王…③王…　Ⅲ. 英语—阅读教学—高等学校：技术学校—教材
②英语—写作—高等学校：技术学校—教材　Ⅳ. H31

中国版本图书馆 CIP 数据核字 (2009) 第 154531 号

出　版　人：于春迟
项目负责：王建平　聂海鸿
责任编辑：聂海鸿
封面设计：孙莉明
版式设计：涂　俐
出版发行：外语教学与研究出版社
社　　　址：北京市西三环北路 19 号 (100089)
网　　　址：http://www.fltrp.com
印　　　刷：北京京师印务有限公司
开　　　本：787×1092　1/16
印　　　张：16.25
版　　　次：2009 年 8 月第 1 版　2009 年 8 月第 1 次印刷
书　　　号：ISBN 978 - 7 - 5600 - 8948 - 5
定　　　价：33.90 元 (含 CD-ROM 一张)
＊　　＊　　＊

前　言

　　高职高专教育是我国高等教育的一个重要组成部分，高职高专学生是我国大学生中一个十分重要的群体。针对这一学生群体的特点，教育部于2000年颁布了《高职高专教育英语课程教学基本要求》(试行)。该《基本要求》明确指出，高职高专的英语教学应该以培养学生实际运用语言的能力为目标，突出教学内容的实用性和针对性。

　　根据《基本要求》的这一指导思想，外语教学与研究出版社组织编写了《新起点大学基础英语教程》系列教材。这套教材自2004年出版以来，被众多高职高专院校采用，作为提高学生英语综合应用能力的主要教材，受到师生的广泛好评。近年来，随着我国社会与经济的发展，国家对高职高专院校人才培养提出了更明确的要求，高职高专院校的英语教学改革也在不断深入。面临新的发展和新的要求，《新起点大学基础英语教程》的编者遵循教育部的指导方针，结合实际使用中的反馈意见，经过认真细致的调研、策划与筹备，对教材进行了认真修订，以满足新形势下高职高专英语教学的需求。

　　为配合高职高专院校的教学安排，《新起点大学基础英语教程》(修订版)每册调整为10个单元，删除并更换了部分选篇。其中《读写教程》还重新编写了语法内容，应用英语写作的教学从第一册开始。此外，《读写教程》和《听说教程》还配备了助学光盘，使该套教材更加立体化。

　　《新起点大学基础英语教程》(修订版)设1-4级，供两个学年使用。每一级别均由《读写教程》、《听说教程》、《学习方法与阅读》和《自主综合训练》组成。与教材配套的还有录音带、助学光盘、电子课件和试题库，各院校可根据实际需要选择使用。

编　者
2009年2月

编写说明

本教材是《新起点大学基础英语教程》中的"读写"系列,与"听说"系列和"学习方法与阅读"系列话题融通,技能互补,构成一个整体。

本书共10个单元,每单元由3部分组成,第一部分的教学内容围绕两篇主题相同的阅读文章展开,侧重阅读理解的训练,同时兼顾写作、口语的训练,以及词汇、语法和文化知识的学习。第二部分对本单元阅读课文中所涉及的词汇和短语进行深入讲解。第三部分是写作训练,分为基础写作练习和实用写作练习两部分。

就学习过程而言,每个单元以阅读课文为核心,学习者通过自上而下、从内容到形式、输入与输出的多重反复等学习活动,不断加深对所学技能与知识的掌握。单元各部分内容的主要编写目的是:

项 目			主要编写目的
Text A and Text B	Before Reading		通过问题激发学生的阅读兴趣,激活学生已有的背景知识,为下一步的阅读理解做准备。
	Reading		通过对阅读文章的分析和讲解帮助学生理解课文。
	After Reading	Main Idea	通过概括性的练习帮助学生对课文进行宏观分析,掌握文章的中心思想。
		Detailed Understanding	通过判断对错与填空两种练习形式加深对课文的理解。
		Detailed Study of the Text	通过对难度较大的词、句的分析,以及对相关背景知识的介绍,帮助学生进一步加深对课文的理解。
		Talking About the Text / Further Work on the Text	通过问答的形式,让学生在练习口语的同时检查自己对阅读文章的掌握程度。
		Vocabulary Practice	通过各种练习帮助学生巩固常用高频词的用法,同时练习文章中出现的生词和词组。
Word Study			从词性、词义、常用搭配及派生词等各个角度对本单元的重点单词进行详细分析和讲解,帮助学生掌握重点词汇。
Writing Practice	Exercises I—II/III		语法练习。帮助学生复习和巩固基础语法知识。
	Exercises III/IV—VI/VII		针对本单元的写作和翻译。巩固本单元的重点句型结构和词汇的用法。
	Exercise VII/VIII		应用文写作。帮助学生熟悉应用文的风格,练习应用文的写法。

Contents

Contents

1
Unit

 Text A

 Before Reading

Discuss the following questions in class.

1. Have you been to museums and galleries? If yes, which museum or gallery do you like best?
2. What are the similarities and differences between museums and galleries on the one hand, and colleges and universities on the other hand?

 Reading

Adult Learning in Museums[1] and Galleries[2]

1   Learning in museums and galleries—that's just for school kids, right? Wrong!

2   Museums and galleries can be interesting learning environments for people of all ages, cultures and backgrounds[3]. And interest is the key—on the whole, children are in compulsory[4] education, but there's nothing compulsory about adult learning. If you don't engage adults' interest, they can just quit learning at any time.

3   Adult learning is high on the Government's agenda[5]; recent projects such as the formation[6] of the Learning and Skills Councils[7] to manage all post-16 education and the University for Industry are just two examples of the current trend in learning—away from old style "education". Learning in the 21st century is something that delivers not facts to be digested[8] under controlled conditions, but knowledge that can be applied to novel[9] situations and challenges—to emphasize[10] real understanding and innovation and to equip you to learn "how to learn", and keep learning throughout your life—something that is becoming essential in a world where the only constant[11] is change. It's the speed of change that's driving many adults back into learning—what they knew yesterday doesn't always apply today.

4   The other element[12] driving up the number of adults in learning has to do with our population (for example, the number of older people will have doubled from 6 million to 12 million between 1961 and 2021). Vitally[13] for museums and galleries, policy-makers are now becoming more interested in learning outcomes[14], rather than who delivers[15] it, which means the boundaries[16] between formal[17] providers (like colleges) and non-formal providers[18] (like museums and galleries) are breaking up[19] as the desire to give people access[20] to learning whenever and wherever they are increases.

5   Museums and galleries are perfectly placed to rise to the challenge and become one of the new types of flexible adult learning suppliers[21]. They have rich resources, are rooted in their communities and have the power to motivate[22] learning and build the confidence of adult learners. Take a recent project in Oxfordshire as an example. Drawn from Memory was funded by the Department of Education and Skills and saw older people participate in creative writing sessions[23] using museum exhibits[24]. One session introduced participants to create calendars from their own old photos of Oxfordshire. Most had never used computers before but with the motivation of the end product (the calendar with their own photos) in mind, they overcame their fears by working in a supportive[25] environment, with like-minded people. Participants commented afterwards, "They'll never believe what I've done!"

6   We would like to see a day when museums and galleries are seen as a first point of call for adult learning by providers and learners alike. We still have a way to go, but if we act now, we can reach this goal and help to change people's lives.

New Words and Expressions

1   museum /mjuːˈzɪəm/ n. 博物馆
2   gallery▲ /ˈgælərɪ/ n. 美术馆，画廊
3   background /ˈbækgraʊnd/ n. 背景，经历
4   compulsory■ /kəmˈpʌlsərɪ/ a. 强制性的，必须做的
5   agenda★ /əˈdʒendə/ n. 议事日程
6   formation▲ /fɔːˈmeɪʃən/ n. 形成，组成
7   council★ /ˈkaʊnsəl/ n. 委员会，理事会
8   digest★ /dɪˈdʒest/ vt. 吸收，领悟
9   novel /ˈnɒvəl/ a. 新的；新颖的
10   emphasize /ˈemfəsaɪz/ v. 强调，着重
11   constant /ˈkɒnstənt/ n. 恒定的事物，不变的事
12   element /ˈelɪmənt/ n. 成分，要素
13   vitally /ˈvaɪtəlɪ/ ad. 重大地；紧要地
     vital /ˈvaɪtəl/ a. 重大的；紧要的
14   outcome /ˈaʊtkʌm/ n. 结果
15   deliver /dɪˈlɪvə(r)/ v. 提供（服务）

16 boundary ★ /'baʊndərɪ/ *n.* 分界线，界限
17 formal /'fɔːməl/ *a.* 正式的，正规的
18 provider# /prəʊ'vaɪdə(r)/ *n.* 供给者
19 break up 消散
20 access /'ækses/ *n.* 享用权，享用机会
21 supplier# /sə'plaɪə(r)/ *n.* 供应者，供给者
22 motivate ★ /'məʊtɪveɪt/ *vt.* 激励，激发
23 session ▲ /'seʃən/ *n.* （从事某项活动的）一段时间
24 exhibit /ɪg'zɪbɪt/ *n.* 陈列品；展品
25 supportive# /sə'pɔːtɪv/ *a.* 支持的；赞许的

Proper Noun

Oxfordshire /'ɒksfədʃɪə(r)/ 牛津郡 [英国英格兰郡名]

After Reading

A. Main Idea

Complete the following diagram with the sentences or expressions given below.

1. have rich resources
2. In today's world, the only constant is change.
3. have the power to motivate learning and build the confidence of adult learners
4. preferable learning providers
5. can be interesting learning environments for all kinds of people
6. are rooted in the learners' communities
7. The population of the older people is increasing rapidly.
8. the Learning and Skills Councils
9. the University for Industry
10. Museums and galleries can be regarded as a first point of call for adult learning by providers and learners alike.
11. the Drawn from Memory project in Oxfordshire

The Main Topic → Museums and galleries _____ _____ _____ .

Adult learning is high on the Government's agenda. → Two recent projects:
a. _____ _____
b. _____ _____

Two elements that make it necessary for adults to keep learning.
a. _____ _____

b. _____ _____

Why are museums and galleries _____ _____ ?
Museums and galleries
a. _____ ,
b. _____ ,
a. and _____ ,
_____ _____ .

A good example: _____ _____

The Author's Hope:
_____ _____ _____ _____

B. Detailed Understanding

I. **Tell if the following statements are true (T) or false (F) according to the text.**

1. _____ People often mistakenly regard learning in museums and galleries as only for school kids.

2. _____ Children have to go to school, and so do adults.

3. _____ The Government is mainly interested in improving higher education.

4. _____ The Government formed the Learning and Skills Councils and the University for Industry to help create more chances for adults to learn.

5. _____ In the 21st century, people don't have to learn any facts.

6. _____ Adults in the 21st century have to learn how to learn.

7. _____ People have to learn throughout their lives because things are changing too rapidly.

8. _____ Policy-makers used to be interested in education outcomes rather than education providers.

9. _____ The learning provided by museums and galleries is more flexible than that of schools and universities.

10. _____ Museums and galleries can motivate people to learn.

11. _____ Older people on the Drawn from Memory project used computers to make calendars from their own photos.

12. _____ The older participants were proud of what they had made.

II. **Explain the sentences by filling in the blanks.**

1. **Text sentence:** ... but there's nothing compulsory about adult learning.
   Interpretation: ... but adults _____.

2. **Text sentence:** If you don't engage adults' interest, they can just quit learning at any time.
   Interpretation: If you fail _____, adults may stop learning at any time.

3. **Text sentence:** Adult learning is high on the Government's agenda...
   Interpretation: The Government gives _____...

4. **Text sentence:** Learning in the 21st century is something that delivers not facts to be digested under controlled conditions, but knowledge that can be applied to novel situations and challenges...
   Interpretation: Instead of _____, learning in the 21st century delivers knowledge that can be used in new situations and challenges...

5. **Text sentence:** ... what they knew yesterday doesn't always apply today.

   Interpretation: ... what people learned in the past _____.

6. **Text sentence:** ... the boundaries between formal providers (like colleges) and non-formal providers (like museums and galleries) are breaking up...

   Interpretation: ... there is _____ between formal providers and non-formal providers...

7. **Text sentence:** Museums and galleries are perfectly placed to rise to the challenge and become one of the new types of flexible adult learning suppliers.

   Interpretation: Museums and galleries, as a new type of flexible adult learning supplier, are _____.

8. **Text sentence:** ... but with the motivation of the end product (the calendar with their own photos) in mind, they overcame their fears by working in a supportive environment, with like-minded people.

   Interpretation: ... but keeping in mind _____, they shared the same objective, supported each other and thus overcame their fears.

C. Detailed Study of the Text

1　**And interest is the key—on the whole, children are in compulsory education...** (Para. 2) 关键是兴趣——总的来说，孩子们接受的是义务教育……

很多国家都规定少年儿童有接受教育的义务，如英国规定5到16岁的少年儿童必须接受教育，美国的大部分州也规定少年儿童在16岁前必须接受义务教育，义务教育一般是免费的。

2　**... the formation of the Learning and Skills Councils to manage all post-16 education...** (Para. 3) ……成立了"学习与技能委员会"，负责16岁以后的教育……

16岁以后的教育指非义务教育，通常指职业教育、普通高等教育以及其他形式的成人教育或继续教育等等。

3　**... something that is becoming essential in a world where the only constant is change.** (Para. 3) ……（在21世纪，学习是）某种在一个变化无常的世界里越来越必不可少的东西。

constant在数学和物理学中解释为"常数"或"恒定（值）"。此句中where the only constant is change意为"不变的规律是变化"。

4　... they overcame their fears by working in a supportive environment, with like-minded people. (Para. 5) ……通过与志同道合的人一起工作，相互支持，他们克服了恐惧。
their fears是指他们对电脑的恐惧。

5　We still have a way to go, but if we act now, we can reach this goal and help to change people's lives. (Para. 6) 我们还有很长的路要走，但如果我们从现在做起，就能够实现这一目标，并帮助人们改变生活。
have a way to go等于have a long way to go，意为"还有很长的路要走"或"还有很多的事要做"。this goal指的是前一句中讲的museums and galleries are seen as a first point of call for adult learning by providers and learners alike。

D. Talking About the Text

Work in pairs. Ask and answer the following questions first and then put your answers together to make an oral composition.

1. What can museums and galleries be for all kinds of people?
2. Has the Government realized the importance of museums and galleries?
3. What should people in the 21st century be able to do?
4. What else should they do throughout their lives? Why?
5. What is another reason for people to keep learning?
6. Why is this an important opportunity for museums and galleries?
7. Why can museums and galleries be good adult learning suppliers?
8. What does the author hope providers and learners can do?

E. Vocabulary Practice

I.　Fill in the blanks with the new words or expressions from Text A.

1. I have read the report but have not _____ everything.
2. The electronic age is forcing us to look at ourselves and our brains in a _____ way.
3. She introduced a new issue and _____ its importance to us.
4. An agreement on the _____ of a new government was reached on June 6.
5. Water is a compound containing the _____ of hydrogen and oxygen.
6. He usually speaks casually, but today he gave a _____ speech.

7. When he was asked why he helped the child, he said he was _____ by love, and expected nothing in return.

8. The _____ of the election was out of the expectation of the general public.

9. Anyway, she is a popular figure _____ she is.

10. The insurance industry plays a _____ role in the British economy.

II. Complete the following dialogs with appropriate words or expressions from Text A.

1. A: What's her educational _____?

   B: She's got a bachelor's degree in Physics from Nanjing University.

2. A: Is this a _____ course?

   B: No, it's an elective course. You can decide whether to take it or not.

3. A: Let's move to the last item on the _____.

   B: Yes, sir.

4. A: What _____ her to leave home?

   B: Her parents disagreed with her in almost every aspect of her life.

5. A: Tom and Jane are next-door neighbors.

   B: That's right, and the fence marks the _____ between Tom's property and Jane's.

6. A: We are one of the largest _____ of employment in the area.

   B: That's why so many applicants want to get a job here.

7. A: Hey Mary, I heard that your _____ won the first prize at the flower show. Congratulations!

   B: Thank you very much.

8. A: What do you think are the _____ of a good life?

   B: Honesty, industry and kindness.

9. A: I like milk, but I can't _____ it.

   B: What a pity! Maybe you can try soya bean milk.

10. A: Your ad says that you don't need any _____ qualifications for this job?

    B: Yes, but we would like you to have some experience.

Before Reading

Discuss the following questions in class.

1. When you graduate from college, would you like to go on studying for higher degrees? If yes, how long would you like to study?

2. How do you understand the saying "One is never too old to learn"?

Reading

Something to Rely On

1   Learning has always been an important part of my life. In fact, I have never stopped studying since graduating over 15 years ago. Strangely, the more knowledge I have acquired[1], the more I have found myself lacking[2] in knowledge. It seems that, once the process of learning has been started, it is hard to stop.

2   After obtaining my first degree in mechanical engineering[3], I studied building services[4] and environmental technology[5]. I obtained an honors degree in physics and mathematics, a postgraduate[6] diploma[7] in construction project management[8], an MBA[9] and a Master's in English for Special Purposes and took some other courses without academic awards. Today, I am pursuing a part-time Ph.D.[10] in English Studies.

3   Some friends wonder how I have kept studying for so many years. They usually ask: "Is it worth studying that much?", "What are the motivations[11]?" and "How can you cope with the pressure of working full-time and studying part-time?"

4   I always find it hard to give definite[12] answers. I have never actually weighed[13] the cost of studying a programme against the expected returns and have not had any specific[14] objectives in mind when I enrolled. All I thought was that the programme was interesting and could provide me with the knowledge I wanted. Probably, my appetite for knowledge gave me the motivation to go on.

5   Taking a formal programme helps one acquire in-depth knowledge in a structured way. I see academic awards as milestones[15] of personal achievement.

6   Like many other part-time adult learners, I do sometimes feel the pressure of working full-time and studying part-time. Relaxing myself for short periods whenever

I accomplish[16] a specific task provides relief[17]. But the major trick is to ensure[18] that I have chosen something I really enjoy. If we long to study something and really want to study it well, we have the internal motivation that turns pressure into interest, plus the passion to keep ourselves going.

7   Academic degrees can sometimes create embarrassment and make an unfavourable[19] impression, should our jobs lag[20] far behind our academic achievements. People may think that we are simply "nerds[21]". If you are one of these unlucky people, don't blame[22] yourself too much, because moving up in the workplace[23] and getting academic achievements are very different games. The success of the former depends much more on luck and human skills than on intelligence[24].

8   Looking on the bright side, I see studying and achieving academic awards as a road to realize my personal dreams—a road on which no one will hold me back. Compared with the returns we get in our jobs and emotional lives, academic achievements are more controllable[25] and predictable[26]. So, some day, I hope to present papers in international conferences[27], see my articles published in journals and even write a book!

New Words and Expressions

1   acquire ★ /əˈkwaɪə(r)/ vt. 得到
2   lack /læk/ v. 缺乏
3   mechanical engineering 机械工程学
4   building service 建筑服务
5   environmental technology 环境技术
6   postgraduate# /ˌpəʊstˈgrædjuət/ n. 研究生
7   diploma ■ /dɪˈpləʊmə/ n. 毕业文凭
8   construction project management 建筑工程管理
9   MBA=Master of Business Administration 工商管理硕士
10   Ph.D.=Doctor of Philosophy 博士
11   motivation /ˌməʊtɪˈveɪʃən/ n. 动机
12   definite /ˈdefɪnɪt/ a. 明确的
13   weigh /weɪ/ v. 权衡
14   specific /spɪˈsɪfɪk/ a. 具体的
15   milestone# /ˈmaɪlstəʊn/ n. 里程碑

16 accomplish * /əˈkʌmplɪʃ/ v. 达到（目的），完成（任务）
17 relief /rɪˈliːf/ n. 轻松，宽慰
18 ensure /ɪnˈʃʊə(r)/ vt. 确保，保证，担保
19 unfavourable# /ˌʌnˈfeɪvərəbl/ a. 不利的；不好的
20 lag * /læg/ vi. 落后，走得慢
21 nerd■ /nɜːd/ n. 乏味的人，呆子
22 blame /bleɪm/ vt. 指责，责怪
23 workplace# /ˈwɜːkpleɪs/ n. 工作场所
24 intelligence * /ɪnˈtelɪdʒəns/ n. 智力，智慧，理解力
25 controllable# /kənˈtrəʊləbl/ a. 可控制的；可支配的
26 predict /prɪˈdɪkt/ v. 预知，预言
    predictable /prɪˈdɪktəbl/ a. 可预言的；可预报的
27 conference * /ˈkɒnfərəns/ n.（正式）会议

 # After Reading

A. Main Idea

Complete the following diagram with the sentences or expressions given below.

1. to present papers in international conferences, publish articles in academic journals
2. to relax myself for short periods whenever I accomplish a specific task
3. to ensure that I have chosen something I really enjoy
4. The more knowledge I have acquired, the more I have found myself lacking in knowledge.
5. workplace success
6. I wanted to get any returns
7. why I have kept studying for so many years
8. I had any specific objectives in mind
9. attended different programmes and obtained different diplomas and degrees
10. much more on luck and human skills than on intelligence
11. my appetite for knowledge gave me the motivation to go on

One thing strange about learning:

_____

_____

_____

As a result, I _____

_____

_____ .

Some friends ask me

_____

_____

_____

_____

I did not study because

_____

_____ or because

_____

_____ .

I kept on learning because

_____

_____

_____ .

Learning—An Important Part of My Life

One problem I have is that I sometimes feel the pressure of working full-time and studying part-time.

One solution is _____

_____

_____ .

Another solution is _____

_____

_____ .

The relationship between academic degrees and workplace success:

More academic degrees may not necessarily mean

_____ .

Moving up in the workplace depends _____

_____

_____ .

I see studying and achieving academic awards as a road to realize my personal dreams.

I hope _____

_____

_____

and even write a book.

13

B. Detailed Understanding

I. Make correct statements according to the text by combining appropriate sentence parts in Column A with those in Column B.

Column A	Column B
1. The more knowledge I have acquired, _____.	a. once it has been started
2. I can hardly stop the process of learning _____.	b. to work full-time and study part-time
3. I have obtained various diplomas and degrees _____.	c. the cost and benefit of my learning
4. Some friends don't understand _____ _____.	d. what kind of returns we can get in our jobs and emotional lives
5. I have never thought about _____ _____.	e. simply because I was interested in it
6. I wanted to learn something _____ _____.	f. why I have kept learning for so many years
7. I also find it hard _____.	g. can turn pressure into interest
8. The internal motivation _____.	h. mainly depends on intelligence
9. Having too many academic degrees _____.	i. in science, engineering, business and liberal arts
10. Getting academic achievements _____.	j. sometimes may make us look like "nerds"
11. It is hard to predict _____.	k. the more knowledge I need

II. Explain the sentences by filling in the blanks.

1. **Text sentence:** Strangely, the more knowledge I have acquired, the more I have found myself lacking in knowledge.

   Interpretation: The strange thing is that the more I have learned, _____.

2. **Text sentence:** Is it worth studying that much?

   Interpretation: Is it worthwhile _____?

3. **Text sentence:** I have never actually weighed the cost of studying a programme against the expected returns...

   Interpretation: I have never compared _____ with _____...

4. **Text sentence:** Taking a formal programme helps one acquire in-depth knowledge in a structured way.

   Interpretation: If one takes a formal programme, _____...

5. **Text sentence:** Academic degrees can sometimes create embarrassment and make an unfavourable impression...

   Interpretation: Academic degrees can sometimes make us _____...

6. **Text sentence:** ... should our jobs lag far behind our academic achievements.

   Interpretation: ... if our work _____.

7. **Text sentence:** ... moving up in the workplace and getting academic achievements are very different games.

   Interpretation: ... having success in the workplace requires _____.

8. **Text sentence:** I see studying and achieving academic awards as a road to realize my personal dreams—a road on which no one will hold me back.

   Interpretation: I regard studying and achieving academic success _____.

9. **Text sentence:** Compared with the returns we get in our jobs and emotional lives, academic achievements are more controllable and predictable.

   Interpretation: It is easier to control and predict the academic achievements we make but _____.

C. Detailed Study of the Text

1 **I obtained an honours degree in physics and mathematics...** (Para. 2) 我获得过物理学和数学的优等学位……

在英国和受英国影响的爱尔兰、印度、新加坡等国的高等教育制度中，一般的本科毕业生可获得合格学位（pass degree），优秀毕业生可获得优等学位（honours degree）。优等学位又分一级、二级和三级（first, second, and third class）。

2 **I have never actually weighed the cost of studying a programme against the expected returns...** （Para. 4）我实际上从来没有将学习的付出与期望的回报相比较……

weigh A against B 意为"权衡 A 与 B",如：weigh one plan against another 权衡一个计划与另一个计划的优劣。

3   **Relaxing myself for short periods whenever I accomplish a specific task provides relief.** (Para. 6) 每当我完成了一项任务，便短时间放松一下，这样可以解除压力。

从 relaxing 到 a specific task 这一部分为动名词短语，作句子的主语。

4   **Academic degrees can sometimes create embarrassment and make an unfavourable impression, should our jobs lag far behind our academic achievements.** (Para. 7) 如果我们的工作成就远远不及我们的学习成绩，那么学位有时会带来尴尬，或给人留下不好的印象。

后半句是 should 引导的一个条件状语从句，等同于 if our jobs lag...。lag behind 意为"落后（于）"，如：The wounded soldiers lagged far behind. 伤员们远远地落在后面。/lag behind other countries in high technology 在高科技方面落后于其他国家。

5   **The success of the former depends much more on luck and human skills than on intelligence.** (Para. 7) 前者的成功更多地取决于运气与人际关系，而非智力。

句中 more... than... 表示对两者中后者的否定。

D. Further Work on the Text

Write down at least three more comprehension questions of your own. Work in pairs and ask each other these questions. If you can't answer any of these questions, ask your classmates or the teacher for help.

1. _____

2. _____

3. _____

E. Vocabulary Practice

I. **Find the word that does NOT belong to each group.**

1. A. postgraduate   B. milestone   C. workplace   D. construction

2. A. management   B. impression   C. achievement   D. embarrassment

3. A. engineering   B. physics   C. mathematics   D. programme

4. A. construction   B. management   C. environmental   D. personal

5. A. MA   B. NBA   C. Ph.D.   D. MBA

II. **Complete the following sentences with appropriate words in their correct form.**

1. **strange, strangely, stranger**

    1) The whole evening seemed _____ unreal.

    2) Do you know why she is so _____ to him?

    3) His habitual place at the table was occupied by a _____.

2. **manage, manager, management**

    1) The farm prospered through good _____.

    2) In spite of these insults (污辱), she _____ not to get angry.

    3) A _____ has to learn some economics if he wants to improve his _____.

3. **construction, construct, constructive**

    1) Most of the factories under _____ are designed by Chinese engineers.

    2) John made a number of _____ suggestions at the meeting.

    3) It takes about two years to _____ a bridge.

4. **predict, predictable, prediction**

    1) Some fortune-tellers say that they can _____ future events.

    2) His _____ seldom came true.

    3) The snow had a _____ effect on traffic.

5. **luck, lucky, luckily, unlucky**

    1) She was _____ to catch a cold on the first day of her holiday.

    2) If you search the site again, you might be _____ enough to find the missing bar.

    3) In the Changchun Book Trade Fair of the year 2000, we ran into him _____.

    4) He came to Beijing to try his _____.

Word Study

post

v.　1. 贴出：They posted the notice at the station. 他们在车站贴出通告。

　　2. 宣布；公告：The ship was posted as missing. 这艘船已宣布失踪。

　　3. 投寄：They will post me the tickets as soon as they receive my check. 他们一收到我的支票就会把票寄给我。/Could you post this letter for me? 你能替我寄这封信吗?

n. 1.（支）柱；标杆：The dog was chained to a post outside. 狗被链子拴在外面的柱子上。

2. 邮政：I'll send the package by post. 我要把这个包裹邮寄出去。

3. 职位；岗位：She is well qualified for the post. 她完全有资格担任这一职位。

派　postage, postal

current

n. 1.（空气、水等的）流；潮流；流速：The current is strongest in the middle of the river. 河中央的水流最湍急。

2. 电流：Turn on the current. 接通电源。

a. 1. 现时的；当前的：current problems 当前的问题/the current issue of a magazine 最新的一期杂志

2. 通行的；流行的：That word is no longer in current use. 那个词已不通用了。

派　currency

access

n. 1. 接近；进入：We gained access into the house through the window. 我们从窗户进入屋内。

2. 通道：The only access to that room is along this hallway. 到那个房间的唯一通道是这条走廊。

have access to 有……的机会；有……权利：Every student has free access to the library. 每个学生都可以免费使用这个图书馆。

lack

v. 缺乏；不足；没有：He failed because he lacked confidence in himself. 他因缺乏自信而失败了。

n. 缺乏；不足；没有：He felt tired for lack of sleep. 他因缺少睡眠而感到疲劳。

pursue

v. 1. 追赶；追踪：The police pursued the stolen vehicle along the highway. 警察沿着公路追赶那辆被盗的车辆。

2. 追求；从事：Economic growth should not be pursued at the expense of environmental pollution. 寻求经济发展不能以污染环境为代价。/He has been pursuing the study of physics for nearly 30 years. 他从事物理学研究已近 30 年了。

派　pursuit

Writing Practice

I. Fill in the blanks with the correct form of the words in brackets.

1. _____ (apply) for the job are required to fill in a form outlining their experience.
2. I find your _____ (behave) quite intolerable.
3. He didn't get the job because he lacked sufficient _____ (confident).
4. This is the _____ (taste) pizza I have ever had.
5. In some villages, the _____ (inhabit) have to walk three miles to get water.
6. I'm afraid 4:30 is _____ (convenient) for me. Can we meet at 5:30 instead?
7. My job is very _____ (stress) and I'm considering quitting.
8. People think he's stupid, but _____ (actual) he's quite intelligent.
9. I don't like learning languages, but generally _____ (speak), I enjoy my English classes.
10. We _____ (agree) about where to go for our holiday, and so ended up staying at home.
11. Her first novel was _____ (incredible) successful, and she became a household name almost overnight.
12. I don't think your behavior is at all _____ (reason) and you ought to apologize to him.
13. The only _____ (advantage) of staying here is that you might get bored.
14. The manager was forced to dismiss 15 of his _____ (employ).
15. He had always hated _____ (fly) and preferred traveling by car.
16. The dog looked dangerous but in fact it was quite _____ (harm).
17. Her _____ (injure) weren't very serious and didn't need hospital treatment.
18. Her job is to look after the _____ (safe) and comfort of the passengers.
19. He was rather _____ (fit) and had to drop out of the race.
20. No _____ (evident) was found to prove the existence of life on other planets.

II. Rewrite the sentences according to the models.

Model A:

Original sentence: Adults can quit learning at any time unless you engage their interest.

   New sentence: If you don't engage adults' interest, they can quit learning at any time. (Text A)

1. Crime rate can rise at any time unless the government takes strict measures.

   _____

2. Your grades can fall at any time unless you work really hard.

   _____

Model B:

Original sentence: The speed of change drives many adults back into learning.

  New sentence: It's the speed of change that's driving many adults back into learning. (Text A)

3. The application of computer technology helps people do things in a more efficient way.

_____

4. The hot-air balloon helped him realize his dream of flying.

_____

Model C:

Original sentence: I only thought that the programme was interesting.

  New sentence: All I thought was that the programme was interesting. (Text B)

5. I only said that he might not come.

_____

6. I only saw the departing train when I got to the station.

_____

Model D:

Original sentence: If one takes a formal programme, it can help him acquire in-depth knowledge in a structured way.

  New sentence: Taking a formal programme helps one acquire in-depth knowledge in a structured way. (Text B)

7. If you practise for two hours a day, it can enhance your performance greatly.

_____

8. If you have a healthy diet, it can help you keep fit until very old age.

_____

III. Combine each set of the sentences into one, using the connective words or expressions provided.

1. a. One recent project is the formation of the Learning and Skills Councils to manage all post-16 education.

   b. Another recent project is the formation of the University for Industry.

   c. These projects are just two examples of the current trend in learning.

   d. The trend of learning is away from old style "education".

   New sentence: _____ (such as, and)

       (Text A)

2.   a. The other element drives up the number of adults in learning.

    b. This element has to do with our population.

New sentence: _____ (V+-ing)

        (Text A)

3.   a. On the one hand, the boundaries between formal providers and non-formal providers are breaking up.

    b. On the other hand, the desire to give people access to learning increases.

    c. People can learn whenever and wherever they are.

New sentence: _____ (as)

        (Text A)

4.   a. Most had never used computers before.

    b. They had the motivation of the end product in mind.

    c. They overcame their fears.

    d. They worked in a supportive environment.

    e. They worked with like-minded people.

New sentence: _____ (but, with, by)

        (Text A)

5.   a. I have acquired a lot of knowledge.

    b. I have found myself lacking in knowledge.

New sentence: _____ (the more..., the more...)

        (Text B)

6.   a. We long to study something.

    b. We really want to study it well.

    c. We have the internal motivation.

    d. The internal motivation turns pressure into interest.

New sentence: _____ (if, and, that)

        (Text B)

7.   a. Maybe you are one of these unlucky people.

    b. In this case, don't blame yourself too much.

    c. Moving up in the workplace and getting academic achievements are very different games.

New sentence: _____ (if, because)

        (Text B)

8.   a. We get returns in our jobs and emotional lives.

    b. Academic achievements are also returns.

    c. Academic achievements are more controllable and predictable.

New sentence: _____ (compare with)

        (Text B)

IV. Translate the following sentences into English.

1. 游泳是一项适合所有年龄层次的体育活动。 （suitable）
2. 整整一个晚上，那个孩子在不停地哭。 （keep doing something）
3. 总的来说，孩子的行为与父母如何教育他们有关。 （have to do）
4. 他想到了老师的忠告，将所有练习又查看了一遍。 （with... in mind）
5. 你一旦开始看这本书，就很难停下来。 （once）
6. 她觉得难以将实情告诉他。 （find）
7. 我将最好的朋友看作是自己的兄弟姐妹。 （see... as）
8. 骑自行车和开汽车是完全不同的两码事。 （game）

V. Translate the following paragraph into Chinese.

Unlike children and teenagers, adults have many responsibilities that they must balance against the demands of learning. Because of these responsibilities, adults have barriers (障碍) against participating in learning. Some of these barriers include lack of time, money, confidence, or interest, lack of information about opportunities to learn, and problems with childcare and transportation. The best way to motivate adult learners is simply to enhance their reasons for enrolling and decrease the barriers.

VI. Practical English Writing

Directions: In today's world there are many things for us to learn. But one thing that is important for all of us is that we should learn how to learn. We can learn from each other, because different students have different learning methods. Now discuss in small groups and put some of the information you learn from other students into the following table.

Method of Learning	Description of the Method	Advantages and Disadvantages of the Method

Basing on what you have learned in the discussion and on the information you find elsewhere, write a short passage on the topic of "Learning How to Learn".

Learning How to Learn

Different people learn in different ways. Some people learn more efficiently while others have a hard time learning. In order to learn more efficiently, I think we must have good methods of learning.

First, _____

_____

_____

_____

_____

_____

_____

To conclude, _____

_____

_____

_____

2
Unit

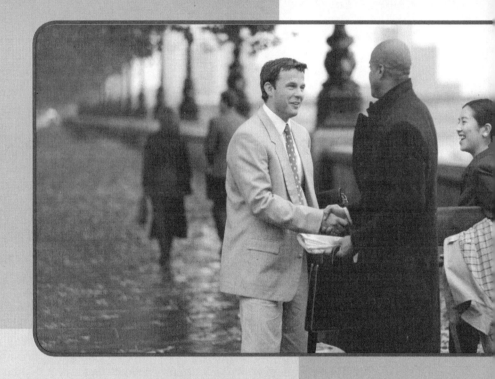

Before Reading

Discuss the following questions in class.

1. Have you ever been misunderstood when communicating with others? If yes, how did you feel when you were misunderstood?

2. Do you want to learn some strategies that can help you avoid being misunderstood by others, especially by people from a different culture? Why?

Reading

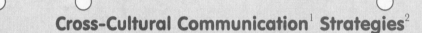

Cross-Cultural Communication[1] Strategies[2]

1   The key to effective cross-cultural communication is knowledge. First, it is essential that people understand the potential[3] problems of cross-cultural communication, and make a conscious[4] effort to overcome these problems. Second, it is important to assume[5] that one's efforts will not always be successful, and adjust[6] one's behavior appropriately[7].

2   For example, one should always assume that there is a significant[8] possibility that cultural differences are causing communication problems, and be willing to be patient and forgiving, rather than hostile[9] and aggressive[10], if problems develop. One should respond slowly and carefully in cross-cultural exchanges, not jumping to the conclusion that you know what is being thought and said.

3   William Ury's suggestion for heated conflicts[11] is to stop, listen, and think when the situation gets tense[12]. By this he means to withdraw[13] from the situation, step back, and reflect[14] on what is going on before you act. This helps in cross-cultural communication as well. When things seem to be going badly, stop or slow down and think. What could be going on here? Is it possible that I misinterpreted[15] what they said, or they misinterpreted me? Often misinterpretation is the source of the problem.

4   Active listening can sometimes be used to check this out—by repeating what one thinks he or she heard, one can confirm that one understands the communication accurately[16]. If words are used differently between languages or cultural groups, however, even active listening can overlook misunderstandings.

5   Often people who are familiar with both cultures can be helpful in cross-cultural communication situations. They can translate both the substance[17] and manner of what is said. For instance, they can tone[18] down strong statements that would be considered appropriate in one culture but not in another, before they are given to people from a culture that does not talk together in such a strong way. They can also adjust the timing[19] of what is said and done. Some cultures move quickly to the point; others talk about other things long enough to establish[20] a relationship with the other person. If discussion on the primary[21] topic begins too soon, the group that needs a "warm-up" first will feel uncomfortable. A person who understands this can explain the problem, and make appropriate adjustments.

6   Yet sometimes the third party can make communication even more difficult. If the third party is of the same culture or nationality[22] as one of the disputants[23], but not the other, this gives the appearance of bias[24], even when none exists. Yet when the third party is of a different cultural group, the potential for cross-cultural misunderstandings increases further. In this case it is necessary to engage in extra discussions about the process and the manner of carrying out the discussions and in confirming[25] understandings at every step in the dialogue or negotiating[26] process.

New Words and Expressions

1   cross-cultural communication 跨文化交际
2   strategy ★ /ˈstrætədʒɪ/ n. 策略
3   potential /pəʊˈtenʃəl/ a. 潜在的，有可能的
4   conscious /ˈkɒnʃəs/ a. 有意识的
5   assume ★ /əˈsjuːm/ vt. 假定，假设
6   adjust /əˈdʒʌst/ v. 调整，调节
7   appropriately# /əˈprəʊprɪətlɪ/ ad. 适当地；恰当地
8   significant /sɪgˈnɪfɪkənt/ a. 重大的
9   hostile ★ /ˈhɒstaɪl/ a. 敌对的，敌意的，不友善的
10   aggressive ▲ /əˈgresɪv/ a. 侵犯的，挑衅的
11   conflict /ˈkɒnflɪkt/ n. 冲突
12   tense /tens/ a. 紧张的
13   withdraw /wɪðˈdrɔː/ v. 撤退
14   reflect /rɪˈflekt/ v. 反思

15 misinterpret# /ˌmɪsɪnˈtɜːprɪt/ v. 误解，误译

    misinterpretation# /ˈmɪsɪnˌtɜːprɪˈteɪʃən/ n. 误解，误译

16 accurately# /ˈækjʊrətlɪ/ ad. 精确地；准确地

17 substance /ˈsʌbstəns/ n. 真正的意义

18 tone★ /təʊn/ vi. （尤指颜色）调和，和谐

    tone down （使）缓和；使协调

19 timing# /ˈtaɪmɪŋ/ n. 时间选择；时机掌握

20 establish /ɪˈstæblɪʃ/ v. 建立

21 primary /ˈpraɪmərɪ/ a. 首要的，主要的

22 nationality /ˌnæʃəˈnælətɪ/ n. 国籍，民族

23 disputant# /dɪsˈpjuːtənt/ n. 争论的一方

    dispute★ /dɪsˈpjuːt/ v. 争论，争吵

24 bias■ /ˈbaɪəs/ n. 偏见，偏心，偏袒

25 confirm /kənˈfɜːm/ v. 进一步确定

26 negotiate★ /nɪˈɡəʊʃɪeɪt/ vt. 洽谈，协商，谈判

Proper Noun

William Ury /ˈwɪljəmˈjʊərɪ/ 威廉·尤里 [人名]

 After Reading

A. Main Idea

Complete the following diagram with the sentences or expressions given below.

1. They can translate both the substance and the manner of what is said.

2. They may further increase misunderstandings if they are from a different culture.

3. They can sometimes make communication more difficult if they are from the same culture as one of the disputants.

4. They can adjust the timing of what is said and done.

5. You should be patient and forgiving, rather than hostile and aggressive.

6. help avoid misinterpretation

7. stop or slow down and think

8. You should respond slowly and carefully instead of jumping to the conclusion.

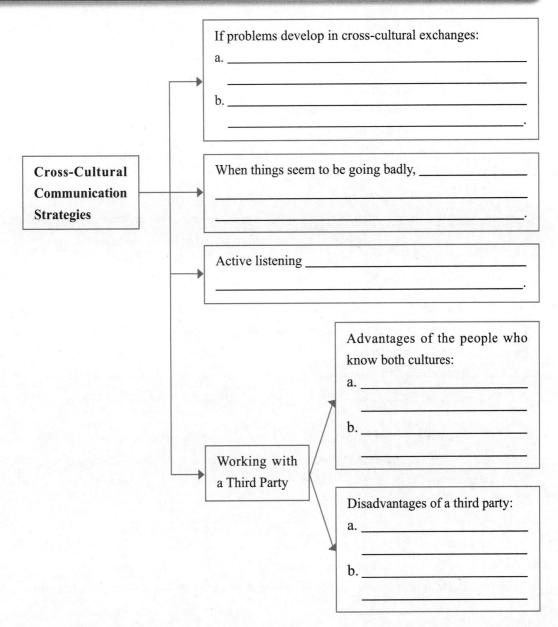

If problems develop in cross-cultural exchanges:

a. _____
_____

b. _____
_____.

Cross-Cultural Communication Strategies

When things seem to be going badly, _____
_____
_____.

Active listening _____
_____.

Advantages of the people who know both cultures:

a. _____
_____

b. _____
_____

Working with a Third Party

Disadvantages of a third party:

a. _____
_____

b. _____
_____

B. Detailed Understanding

I. **Tell if the following statements are true (T) or false (F) according to the text.**

1. _____ The key to effective cross-cultural communication is a potential problem for all people.

2. _____ It is said that people's efforts will not always be successful, and people should adjust their behavior appropriately in cross-cultural communication.

3. _____ Cultural differences hardly cause any communication problems.

4. _____ In cross-cultural communication, people can jump to the conclusion because time is limited.

5. _____ According to William Ury, when a situation gets tense, people should withdraw from the situation, that is to stop, listen and think.

6. _____ William Ury's suggestion was not originally intended for cross-cultural communication.

7. _____ If words are used differently in two different cultures, active listening can prevent misunderstandings.

8. _____ People who know both cultures can be helpful in cross-cultural communication situations.

9. _____ When the third party is of a different cultural group, it will cause bias.

10. _____ The potential for cross-cultural misunderstandings increases further when the third party is from the same culture as one of the disputants.

II. Explain the sentences by filling in the blanks.

1. **Text sentence:** First, it is essential that people understand the potential problems of cross-cultural communication, and make a conscious effort to overcome these problems.

   Interpretation: First, the important thing is that people should understand there are _____.

2. **Text sentence:** One should respond slowly and carefully in cross-cultural exchanges, not jumping to the conclusion that you know what is being thought and said.

   Interpretation: In cross-cultural exchanges one should act slowly and carefully instead of _____.

3. **Text sentence:** By this he means to withdraw from the situation, step back, and reflect on what is going on before you act.

   Interpretation: He means that we should get ourselves out of the situation, and _____.

4. **Text sentence:** This helps in cross-cultural communication as well.

   Interpretation: This also _____.

5. **Text sentence:** Often misinterpretation is the source of the problem.

   Interpretation: It is misinterpretation that _____.

6. **Text sentence:** Active listening can sometimes be used to check this out.

   Interpretation: In order to find out the misinterpretation, people _____.

7. **Text sentence:** If words are used differently between languages or cultural groups, however, even active listening can overlook misunderstandings.

   Interpretation: Even active listening sometimes fails to help you notice misinterpretation if the words _____.

8. **Text sentence:** They can translate both the substance and the manner of what is said.

   Interpretation: They can tell you the actual meaning as well as the way _____.

9. **Text sentence:** They can also adjust the timing of what is said and done.

   Interpretation: They can determine when _____.

10. **Text sentence:** Yet when the third party is of a different cultural group, the potential for cross-cultural misunderstandings increases further.

    Interpretation: When the third party belongs to a different culture, chances are that _____.

C. Detailed Study of the Text

1　**One should respond slowly and carefully in cross-cultural exchanges, not jumping to the conclusion that you know what is being thought and said.** (Para. 2) 在跨文化交流中，应该慢慢地、谨慎地作出反应，而不是仓促地下结论，认为你知道人家在想什么、说什么。句中 jump to the conclusion 意为"仓促地下结论"。

2　**Williams Ury** (Para. 3) 威廉·尤里供职于 Program on Negotiation 大学联合体（简称 PON）。PON 是一家专门研发谈判与争论的理论与实践的机构，总部设在哈佛大学法学院。尤里曾为世界多个政府、企业巨头和社区提供技术支持，现任全球谈判项目组主任。

3　Active listening can sometimes be used to check this out... (Para. 4) 有时，倾听时积极思考便可以发现误解……
   句中 this 指的是上文提到的 misinterpretation。

4　**If the third party is of the same culture or nationality as one of the disputants, but not the other, this gives the appearance of bias, even when none exists.** (Para. 6) 如果第三方与争论人其中一方同属一种文化或来自同一国家，而与另一方无缘，那么即使第三方没有偏见，也会显得有偏见。
   even when none exists 的完整表达应该是：even when there isn't any bias。

D. Talking About the Text

Work in pairs. Ask and answer the following questions first and then put your answers together to make an oral composition.

1. What knowledge should people acquire to carry out effective cross-cultural communication?
2. What should people do if problems develop in cross-cultural communication?
3. In what way is William Ury's suggestion helpful to cross-cultural communication?
4. What good does repetition do to cross-cultural communication?
5. What kind of people may be helpful in cross-cultural communication? In what way can they help?
6. Is the third party helpful?
7. What happens when the third party is from the same culture or nationality as one of the disputants, but not the other?
8. What happens when the third party is of a different cultural group?

E. Vocabulary Practice

I. Fill in the blanks with the new words or expressions from Text A.

1. I didn't want her to jump to the _____ that the divorce was in any way her fault.
2. The driver _____ the policeman's signal and turned in the wrong direction.
3. It may be safely _____ that there is no life on Mars.
4. Sometimes we may _____ other's faults, but never our own.
5. The visitor was given a _____ reception by a crowd of angry farmers.
6. Why have things gone so _____? I've been away for only two days.
7. He will have to make major _____ to his thinking if he is to survive in office.
8. I must ask you to _____ your remarks; they are offending some of our visitors.
9. The government has had to _____ with opposition parties on the new legislation (法规).
10. The author has just finished one book and has another in _____.

II. Complete the following dialogs with appropriate words or expressions from Text A.

1. A: Your house is far from the main road and the bedrooms are small…
   B: Ah yes, we can _____ the rent if you really want to stay here.
2. A: _____ that you got the world champion, what would you do?
   B: I've never dreamed about that.

3. A: A new Pacific defense _____ has been worked out by the new government.

   B: Yes. The safety in the Pacific area has always been at the top of their agenda.

4. A: I don't like her manner—she's very eager to fight.

   B: I agree. Her _____ attitude towards others has offended many people.

5. A: He works slowly but _____.

   B: So he never makes any mistakes.

6. A: What _____ are you，Miss?

   B: French, Sir.

7. A: Excuse me, can you _____ the menu for me?

   B: Yes, sir. But I'm afraid I can't do it word for word.

8. A: After more than 10 hours of flight, I'm feeling tired and having a bad headache.

   B: You'd better make some _____ to the time here.

9. A: I tend to forget most of the things I read.

   B: There is nothing to worry about it. It's a normal part of the learning _____.

10. A: How come? I've missed a whole paragraph! How could I be so stupid?

    B: Oh, no. That's probably because you're too _____.

Before Reading

Discuss the following questions in class.

1. Do you think people from different cultures do business differently?

2. If you want to sell something to a person from a different culture, what do you have to do?

Reading

Culture and International Business

1   Whether we are dealing with issues of marketing, managing, or negotiating, the success or failure of a company abroad[1] depends on how effectively[2] its employees can exercise their skills in a new location. That ability will depend on both their job-related expertise[3] and their sensitivity[4] to the new cultural environment. One of the most common factors contributing[5] to failure in international business assignments is

the misunderstanding that if a person is successful in the home environment, he or she will be equally successful in a different culture.

2   Research has shown that failures in the overseas[6] business setting most frequently[7] result from an inability[8] to understand and adapt to foreign ways of thinking and acting rather than from lacking technical[9] or professional abilities. The literature on international business is filled with examples of conflicts when some corporations attempted to operate in an international context. Some are mildly amusing, others are extremely embarrassing. All of them, to one degree or another, have been costly in terms of money, reputation[10], or both.

3   For example, when American firms try to market their products in other countries, they often assume that if a marketing strategy is effective in Cleveland, it will be equally effective in other parts of the world. But problems can arise[11] in changed cultural contexts. For example, a North Carolina firm purchased a machinery[12] company near Birmingham, England, in hopes of using it to gain entry[13] into the European market. Shortly after the takeover[14], the US manager attempted to correct what he considered to be a major production problem, the time-consuming tea break. The record recounts[15]:

4   In England, tea breaks can take a half-hour per man, as each worker makes his own tea to suit his particular taste and drinks out of a large, pint[16]-size cup. Management[17] suggested to the union[18] that perhaps it could use its good offices[19] to speed up the tea time to ten minutes a break. The union agreed to try but failed. Then one Monday morning, the workers protested[20]. Windows were broken, tomatoes greeted the executives[21] as they entered the plant and police had to be called to keep order. It seems the company went ahead and installed[22] a standard tea vending machine[23] that supplied small paper cups instead of the pint-sized containers, as they are in America. The plant never did get back into production. Even after the tea vending machine was moved out, workers refused to go back to work and the company finally closed down.

New Words and Expressions

1 abroad /əˈbrɔːd/ *ad.* 在国外，到国外
2 effectively# /ɪˈfektɪvlɪ/ *ad.* 有效地
   effective /ɪˈfektɪv/ *a.* 有效的

3　expertise■ /ˌekspɜːˈtiːz/ *n.* 专门知识

4　sensitivity# /ˌsensɪˈtɪvətɪ/ *n.* 敏感性

　　sensitive /ˈsensɪtɪv/ *a.* 敏感的，灵敏的

5　contribute★ /kənˈtrɪbjuːt/ *vi.* (*to*) 是……的部分原因

6　overseas /ˌəʊvəˈsiːz/ *a.* 海外的，国外的

7　frequently /ˈfriːkwəntlɪ/ *ad.* 频繁地

8　inability# /ˌɪnəˈbɪlətɪ/ *n.* 无能力

9　technical /ˈteknɪkəl/ *a.* 技术的，工艺的

10　reputation★ /ˌrepjʊˈteɪʃən/ *n.* 名气，名声

11　arise /əˈraɪz/ *v.* 出现，产生

12　machinery▲ /məˈʃiːnərɪ/ *n.* [总称] 机器，机械

13　entry /ˈentrɪ/ *n.* 进入；入口处

14　takeover# /ˈteɪkˌəʊvə(r)/ *n.* 接管

15　recount# /rɪˈkaʊnt/ *v.* 详述

16　pint▲ /paɪnt/ *n.* 品脱（液量单位）

17　management /ˈmænɪdʒmənt/ *n.* 管理人员

18　union★ /ˈjuːnjən/ *n.* 工会，联盟

19　sb's good offices 某人的帮忙

20　protest★ /prəʊˈtest/ *v.* 抗议，反对

21　executive★ /ɪɡˈzekjʊtɪv/ *n.* 主管，行政人员

22　install★ /ɪnˈstɔːl/ *v.* 安装

23　vending machine *n.* 自动售货机

Proper Nouns

Cleveland /ˈkliːvlənd/ 克利夫兰 [美国俄亥俄州东北部港口城市]

North Carolina /ˌnɔːθˌkærəˈlaɪnə/ 北卡罗来纳州 [美国州名]

Birmingham /ˈbɜːmɪŋəm/ 伯明翰 [英国英格兰中部城市]

 **After Reading**

A. Main Idea

Complete the following diagram with the expressions given below.

1. the inability to understand and adapt to foreign ways of thinking and acting
2. that if a marketing strategy is effective in Cleveland, it will be equally effective in other parts of the world
3. the misunderstanding that if a person is successful in the home environment, he or she will be equally successful in a different culture
4. employees' job-related expertise
5. their sensitivity to the new cultural environment
6. the employees' ability to exercise their skills in a new location
7. the lack of technical or professional abilities

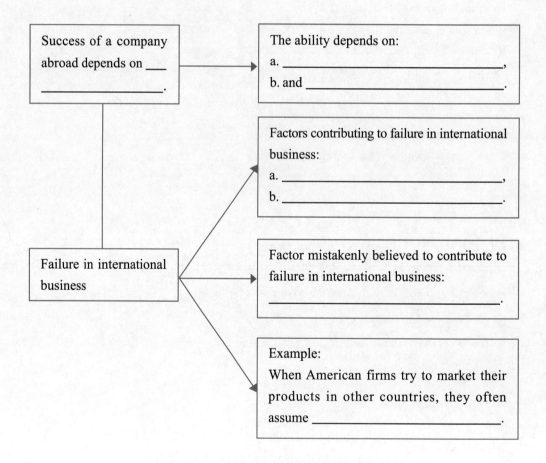

B. Detailed Understanding

I. Make correct statements according to the text by combining appropriate sentence parts in Column A with those in Column B.

Column A	Column B
1. The ability to exercise their skills in a new location will _____.	a. will cause failures in running an overseas business
2. If a person is successful in one's own country, _____.	b. if a marketing strategy is successful in Cleveland, it will be equally successful elsewhere
3. Not being able to understand and adapt to foreign ways of thinking and acting _____.	c. they will face problems
4. All of the examples of conflicts _____.	d. while the British prefer to make tea themselves
5. Some Americans wrongly believe that _____.	e. for the purpose of getting entry into the European market
6. If people move to another culture _____.	f. depend on the job-related knowledge and the sensitivity to a new location
7. A North Carolina firm bought a machinery company near Birmingham _____.	g. it tried to shorten the time-consuming tea break
8. After the US firm took over the machinery company in England _____.	h. have been costly in terms of money, reputation, or both
9. Though the tea vending machine was finally moved out, _____.	i. the workers refused to go back to work
10. Americans like to get tea from tea vending machines _____.	j. he or she may not be successful in another culture

II. Explain the sentences by filling in the blanks.

1. **Text sentence:** That ability will depend on both their job-related expertise and their sensitivity to the new cultural environment.

   Interpretation: That ability is determined by how much they know their job and _____.

2. **Text sentence:** One of the most common factors contributing to failure in international business assignments is the misunderstanding...

   Interpretation: Misunderstanding is a very common factor that leads _____...

3. **Text sentence:** ... if a person is successful in the home environment, he or she will be equally successful in a different culture.

   Interpretation: ... success in a different culture will depend on _____.

4. **Text sentence:** ... failures in the overseas business setting most frequently result from an inability to understand and adapt to... rather than from lacking technical or professional abilities.

   Interpretation: ... it is the inability to understand and adapt to..., but not the lack of technical and professional ability, that causes _____.

5. **Text sentence:** All of them, to one degree or another, have been costly in terms of money, reputation, or both.

   Interpretation: To some degree, all of them cost _____.

6. **Text sentence:** ... if a marketing strategy is effective in Cleveland, it will be equally effective in other parts of the world.

   Interpretation: ... if a marketing strategy works well in Cleveland, it will also _____.

7. **Text sentence:** But problems can arise in changed cultural contexts.

   Interpretation: But when in another culture, _____.

8. **Text sentence:** ... each worker makes his own tea to suit his particular taste and drinks out of a large, pint-size cup.

   Interpretation: ... each worker makes tea according to his own taste and drinks it _____.

9. **Text sentence:** ... it could use its good offices to speed up the tea time to ten minutes a break.

   Interpretation: ... it could make use of its influence to shorten _____.

10. **Text sentence:** ... tomatoes greeted the executives as they entered the plant.

    Interpretation: ... _____ as the executives entered the plant.

C. Detailed Study of the Text

1　**...how effectively its employees can exercise their skills in a new location.** (Para. 1) 它的雇员是如何有效地在新的地方运用他们的技能。

句中 exercise 为动词，表示"运用，使用"的意思。

2　**The literature on international business is filled with examples of conflicts...** (Para. 2) 国际商务文献中有大量关于……冲突的事例。

句中 literature 是"文献"的意思。

3　**All of them, to one degree or another, have been costly in terms of money, reputation, or both.** (Para. 2) 在一定程度上，所有这些都（给企业）在经济或名誉上、或在两方面同时造成了损失。

to one degree or another 等于 to a degree, to a certain degree, to some degree。另外，句中 in terms of 意为"从……方面"。

4　**...tomatoes greeted the executives as they entered the plant...** (Para. 4) 当公司管理人员进入工厂时，西红柿迎头飞来。

这是拟人化的写作手法，类似句子还有：Everything around smiled at him. 或 The rain kept tapping on the windowpanes with a million nervous fingers.

5　**tea time** (Para. 4) 喝下午茶时间。

喝下午茶是英国人传统的生活习惯。传统的下午茶时间大约在下午 4 点。在英国，18 世纪即开始形成饮茶的风气，并沿袭至今。没有人愿意错过享受下午茶的时间。

D. Further Work on the Text

Write down at least three more comprehension questions of your own. Work in pairs and ask each other these questions. If you can't answer any of these questions, ask your classmates or the teacher for help.

1. _____

2. _____

3. _____

E. Vocabulary Practice

I. Find the word that does NOT belong to each group.

1. A. purchase   B. consume   C. effective   D. produce

2. A. failure   B. production   C. reputation   D. negotiation

3. A. understand   B. interpret   C. knowledge   D. communicate

4. A. kind   B. aggressive   C. patient   D. forgiving

5. A. international   B. job-related   C. abroad   D. patient

II. Complete the following sentences with appropriate words in their correct form.

1. **operate, operator, operation, operational**

   1) How well does your new decision-making system _____ in practice?

   2) The college has been in _____ for 90 years now.

   3) Wait, I'll get the _____ to put you through.

   4) The new stadium is fully _____ and ready for business.

2. **amuse, amusing, amusement**

   1) These rural _____ attracted visitors from big cities.

   2) The boy _____ himself in class by drawing a cartoon picture of his math teacher.

   3) His interest was making films for _____.

   4) John, can you drop in? I have some _____ news for you.

3. **profession, professional, professor**

   1) Ted is invited to Georgia University as a visiting _____.

   2) Hurd signed with Real Madrid as a _____ football player.

   3) After college he chose to sell insurance by _____.

4. **consider, considerate, consideration, inconsiderate**

   1) Please give careful _____ to the issue before you make a decision.

   2) I've got a serious suggestion to make, and I want you to _____ it very carefully.

   3) It was very _____ of you to send me a get-well card.

   4) It was _____ of you to arrive unexpectedly.

5. **understand, understanding, misunderstanding**

   1) People who speak the same language can easily _____ one another.

   2) Before the end of the conference I don't want to leave any _____ in your mind.

   3) Unless the two countries soon reach an _____, there will be a war.

Word Study

exercise

vt. 1. 运用（权力）；施加（影响）：I intended to exercise my right to vote the bill down. 我打算对提案行使我的否决权。

2. 锻炼（身体的某一部分）：Climbing exercises all the major muscles in a body. 爬山使身体的主要肌肉得到锻炼。

vi. 运动，锻炼：You are losing your good health; you should exercise. 你的身体不如以前了，该多运动运动。

n. 1. 运动，锻炼：She doesn't have enough exercise. 她缺乏锻炼。

2. 训练，体操：The doctor gave me some exercises to help with my legs. 医生叫我做一些训练来增强腿部力量。

3. 练习，作业：Please turn to Page 123 and do Exercise 1. 翻到第 123 页，做练习一。

do exercises 做操

exist

vi. 1. 存在：Stop pretending that the problem doesn't exist. 别再假装这个问题不存在了。

2. 活着；（on）靠……生存：The hostages existed on bread and water for over seven months. 人质靠面包和水生存了 7 个多月。

派 existence

withdraw

vi. 1. （*from*）退出，不参加：There are calls for Britain to withdraw from the European Union. 有呼声要求英国退出欧盟。

2. 离开：After dinner, we withdrew to the garden for a private talk. 饭后我们退到花园进行密谈。

vt. 1. （*from/out of*）提取（银行存款）：She withdrew all her money from her account. 她提取了账户上所有的钱。

2. 撤回；取消：The UN decided to withdraw all its support for that government. 联合国决定撤销对那个政府的一切支持。

派 withdrawal

heat

n.　1. 热度，热：When the oven reaches the correct heat, the light goes off. 烤箱达到合适的温度时，灯就会熄灭。/Black surface absorbs heat. 黑色的表面吸热。

　　 2. 暑热；暖气：The game went on in the heat of the day. 比赛在一天中最热的时候继续进行着。/The cottage had no heat or water. 这个小村舍没有供暖和供水系统。

　　 3. 热烈，激烈：In the heat of the argument I declared that I stood on her side. 在争论最激烈的时候我宣布我支持她那一方。

vt.　1. 加热：I heated the remains of the lunch. 我把中午剩下的饭菜热了。

　　 2. 激化，加剧：Things were heating up in the North, so I should return immediately. 北方的局势紧急，我得立即返回。

　　 in the heat of 在……的鼎盛时期

派　heated

manner

n.　1. 方式，做法：Please speak in a frank manner. 请坦率地说。

　　 2. 态度；举止：I don't like his cold manner. 我不喜欢他冷漠的态度。

　　 3. 礼貌；规矩：Mind your manners in public places. 在公共场所要注意礼貌。

　　 4. 风俗，习惯：a lecture on the life and manners of Victorian London 一场关于维多利亚时期伦敦生活和风俗的讲座

Writing Practice

I. Complete the phrasal verb in each of the following sentences with an appropriate preposition or adverb.

1. We were out walking when we came _____ a wonderful restaurant.

2. The other team were very good, and as a result we came _____ badly in the game.

3. As I hadn't paid my bill, the electric company cut me _____.

4. Don't cut _____ when I'm speaking—I hate it when people interrupt me.

5. She didn't remarry because she never got _____ the death of her husband.

6. That shop is giving _____ a free printer with every computer they sell.

7. The fire alarm went _____ in the middle of the night.

8. He went _____ working despite the noise next door.

9. Things have been bad for a while, but at last they're beginning to look _____.

10. Look _____! You're driving too fast.

11. I can't make _____ why she didn't come to the party. Maybe she was ill.

12. I don't earn much, but I always manage to put some money _____ at the end of the month.

13. I'm trying to concentrate—please don't put me _____.

14. We've set _____ a company that produces computer parts.

15. He was going to have a long journey, so he had to set _____ before daybreak.

16. I needed to get fit, so I took _____ jogging and swimming in my free time.

17. He offered me a job, but unfortunately I had to turn _____ his offer.

18. I've lost my sunglasses, but they will probably turn _____ somewhere.

19. Thieves broke _____ the bank and stole over 10 million dollars.

20. The car has broken _____ again; we'll have to take it to the repair shop.

II. **Choose the answer which is closest in meaning to the underlined part in each of the following sentences.**

1. The dog underlined:went for him.
   a. attacked
   b. attracted
   c. attached

2. The judges picked his story as the best.
   a. noticed
   b. chose
   c. found

3. I feel for you.
   a. know
   b. find for
   c. sympathize with

4. He called on her.
   a. gave a phone call
   b. visited
   c. talked with

5. He launched into a long speech.
   a. began
   b. wrote
   c. completed

6. The logo stands for the company.
   a. stands in front of
   b. symbolizes
   c. becomes part of

7. They called the trip off.
   a. declared
   b. cancelled
   c. started

8. They laid on a good meal.
   a. provided
   b. went to
   c. planned

9. He jumped at the idea.
   a. was mad at
   b. abandoned
   c. was enthusiastic about

10. She put the fire out.
    a. started
    b. stopped
    c. moved

III. Rewrite the sentences according to the models.

Model A:

Original sentence: One should always be willing to be patient and forgiving, instead of being hostile and aggressive.

New sentence: One should be willing to be patient and forgiving, rather than hostile and aggressive. (Text A)

1. One should always be willing to be kind and helpful, instead of being cold and indifferent.

   _____

2. One should always be hardworking, and should not be lazy.

   _____

Model B:

Original sentence: If a person understands this, he can explain the problem and make appropriate adjustments.

New sentence: A person who understands this can explain the problem, and make appropriate adjustments. (Text A)

3. If a person has attended our program for one week, he can express some basic ideas in English easily.

   _____

4. When a person is employed for the position, he can get one month training in the first year and a two-week vacation every year.

   _____

Model C:

Original sentence: ... a North Carolina firm purchased a machinery company near Birmingham, England, and it hoped to use it to gain entry into the European market.

New sentence: ... a North Carolina firm purchased a machinery company near Birmingham, England, in hopes of using it to gain entry into the European market. (Text B)

5. Mary replied to the e-mail from the president immediately. She hoped to get the scholarship.

   _____

6. He tried an early operation, and hoped to get rid of the cancer.

   _____

Model D:

Original sentence: The union agreed to try, but they didn't succeed.

New sentence: The union agreed to try but failed. (Text B)

7. The committee agreed to start another investigation, but they didn't succeed.

   _____

8. The tutor gave promise of high scores of their child, but she didn't succeed.

   _____

IV. **Combine each set of the sentences into one, using the connective words or expressions provided.**

1. a. William Ury has a suggestion for heated conflicts.

   b. The suggestion is to stop, listen, and think.

   c. The situation gets tense.

   New sentence: _____ (when)

   (Text A)

2. a. Things seem to be going badly.

   b. Stop or slow down and think.

   New sentence: _____ (when)

   (Text A)

3. a. Words are used differently between languages.

   b. Words are used differently between cultural groups.

   c. Even active listening can overlook misunderstandings.

   New sentence: _____ (if, however)

   (Text A)

4. a. Often people are familiar with both cultures.

   b. People can be helpful in cross-cultural communication situations.

   New sentence: _____ (who)

   (Text A)

5. a. The third party is of the same culture or nationality as one disputant.

   b. The third party is not of the same culture or nationality as the other disputant.

   c. This gives the appearance of bias.

   New sentence: _____ (if, but not)

   (Text B)

6. a. Failures in the overseas business setting most frequently have one cause.

   b. People are not able to understand and adapt to foreign ways of thinking and acting.

   c. Research has shown that.

   New sentence: _____ (that)

   (Text B)

7. a. The takeover took place.

   b. The US firm considered something to be a major production problem.

   c. The problem was the time-consuming tea break.

   d. The US firm attempted to correct the problem.

   New sentence: _____ (shortly after, what)

   (Text B)

8. a. The tea vending machine was moved out.

   b. Workers refused to go back to work.

   c. The company finally closed down.

   New sentence: _____ (even after, and)

   (Text B)

V. Translate the following sentences into English.

1. 中国商人正在推销他们的产品，希望获准进入欧洲市场。(in hopes of)
2. 在跨文化交际中，人们应该适时地调整自己的行为，而不应该退避。(adjust, rather than)
3. 我们帮助别人时也帮助了自己。(by doing...)
4. 许多因素导致了这家公司的倒闭。(contribute to)
5. 成年人往往根据自己的经验解释孩子的行为。(interpret, in terms of)
6. 生意的成败取决于一个人与生意相关的知识。(relate)
7. 有些经营方式很有效，有些很耗时。(effective)
8. 他的建议对于解除国际事务中的误解有帮助。(help)

VI. Translate the following paragraph into Chinese.

It's no secret that today's workplace is rapidly becoming vast, as the business environment expands to include various geographic locations and span numerous cultures. What can be difficult, however, is understanding how to communicate effectively with individuals who speak another language or who rely on different means to reach a common goal.

VII. Practical English Writing

Directions: We know that as human beings, we need to communicate with each other for various purposes. In the following table, you can find some of the purposes. Please discuss with your classmates and see if you can find more purposes of communication.

Purpose of Communication	Your Examples	Your Explanation
Establish social connections	Good morning. How do you do?	We greet each other to show friendliness.
Exchange information		
Express emotions		

Now write a short passage entitled "Purposes of Communication" based on the results of your discussion.

Purposes of Communication

3

Unit

Before Reading

Discuss the following questions in class.

1. Do you use the Internet? If yes, how many hours do you spend online every day?

2. What do you think are the problems the Internet may create?

Reading

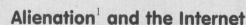

Alienation[1] and the Internet

1   With the increasingly wide use of the Internet, there has been much talk about the "new information age". However, much less widely reported has been the idea that the Internet may be responsible for furthering the splitting[2] up of society by alienating its individual users. At first this might sound like an apparent contradiction[3]: How can something be, on the one hand, responsible for global unification[4] by enabling the free exchange of ideas and, on the other hand, alienating the participants?

2   I had a recent discussion with a friend of mine who has what he described as a "problem" with the Internet. When I questioned him further he said that he was "addicted"[5], and had "forced" himself to go off-line. He went on to say that all of the time that he spent online might have altered[6] his sense of reality, and that it made him feel lonely and depressed.

3   Last weekend my wife and I invited our extended family[7] to our home to celebrate our daughter's birthday. During the celebration my young nephew[8] spent the entire time on my computer playing a virtual[9] war game. My brother-in-law and I were chatting nearby and it struck[10] us that in generations[11] past, his son, my nephew, would have been outside playing with his friends. But now the little fellow goes online to play his games against his friends in cyberspace[12].

4   It seems to me that the Internet is a powerful tool that presents an opportunity for the advancement[13] of the acquisition[14] and application of knowledge. However, based on my personal experience I can understand how, as they surf the Web, some folks might be confronted[15] with cognitive[16] overload. And I can also understand how one might have his or her sense of reality distorted[17] in the process. Is the Internet a real place?

5   Depending upon how a "real place" is defined it might very well be. At the very least, I believe that when we use the Internet, we are forced to ask fundamental[18] questions about how we perceive[19] the world about us—perhaps another unintended consequence. Some would argue that the virtual existences created by some users who debate, shop, travel and have romance online are in fact not real. While others would argue that, since in practical terms, folks are debating, shopping, traveling and having romance, the opposite is true.

6   I am not at all certain where the "information superhighway" will lead us: Some say to utopia[20], while others feel it's the road to hell[21]. But I do know that we all have the ability to maintain our sense of place in the world. Whether we choose to take advantage of this ability is another matter.

New Words and Expressions

1 alienation /ˌeɪljəˈneɪʃən/ *n.* 疏远，离间
   alienate■ /ˈeɪljəneɪt/ *vt.* 使疏远，离间

2 split★ /splɪt/ *vi.* 分裂

3 contradiction★ /ˌkɒntrəˈdɪkʃən/ *n.* 矛盾

4 unification# /ˌjuːnɪfɪˈkeɪʃən/ *n.* 统一，联合

5 addict★ /əˈdɪkt/ *vt.* 使上瘾，使入迷
   addicted /əˈdɪktɪd/ *a.* 上了瘾的，入了迷的

6 alter /ˈɔːltə(r)/ *v.* 改变

7 extended family （包括近亲的）大家庭

8 nephew /ˈnefjuː/ *n.* 侄子，外甥

9 virtual /ˈvɜːtʃʊəl/ *a.* [计] 虚拟的

10 strike★ /straɪk/ *vt.* 使（某人）突然意识到

11 generation★ /ˌdʒenəˈreɪʃən/ *n.* 一代人

12 cyberspace /ˈsaɪbəˌspeɪs/ *n.* 网络空间

13 advancement /ədˈvɑːnsmənt/ *n.* 发展，进步

14 acquisition★ /ˌækwɪˈzɪʃən/ *n.* 获得，习得

15 confront /kənˈfrʌnt/ *v.* 面临

16 cognitive /ˈkɒgnɪtɪv/ *a.* 认知的

17 distort /dɪsˈtɔːt/ *vt.* 扭曲，使变形

18 fundamental★ /ˌfʌndəˈmentəl/ *a.* 基本的，根本的，基础的

19 perceive★ /pəˈsiːv/ *v.* 感知，感觉

20 utopia■ /juːˈtəʊpɪə/ *n.* 乌托邦；理想的完美境界

21 hell▲ /hel/ *n.* 地狱

After Reading

A. Main Idea

Complete the following diagram with the expressions given below.

1. feel lonely and depressed
2. change a user's sense of reality
3. find it hard to go off-line
4. maintain our sense of place in the world
5. virtual existences
6. the distortion of their sense of reality
7. cognitive overload
8. play with friends in cyberspace

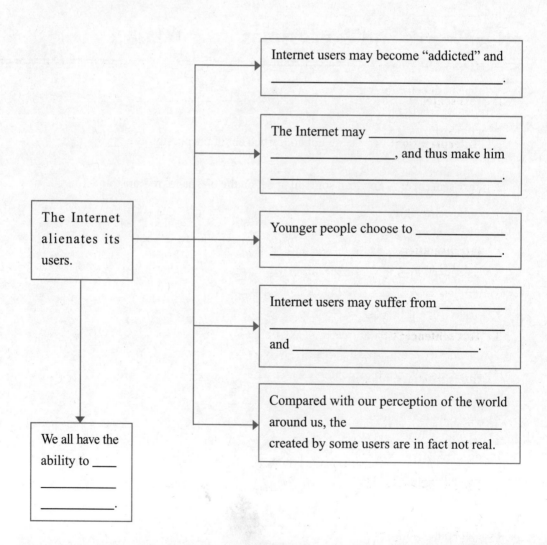

B. Detailed Understanding

I. Tell if the following statements are true (T) or false (F) according to the text.

1. _____ Many people are concerned about the idea that the Internet may be responsible for furthering the splitting up of society by alienating its individual users.

2. _____ Internet users may feel lonely and depressed because the information they get from there may change their sense of reality.

3. _____ Many members of the younger generation become addicted to Internet games.

4. _____ As the Internet contributes to the advancement of the acquisition and application of knowledge, it makes learning easier.

5. _____ The cognitive overload that Internet users are confronted with distorts their sense of reality.

6. _____ The Internet may be regarded as either a real place or an unreal place.

7. _____ The author believes that the information superhighway would lead us to utopia.

8. _____ All Internet users have the ability to maintain their sense of place in the world.

II. Explain the sentences by filling in the blanks.

1. **Text sentence:** However, much less widely reported has been the idea that the Internet may be responsible for furthering the splitting up of society...

   Interpretation: People are not concerned about the possibility _____...

2. **Text sentence:** ... how can something be, on the one hand, responsible for global unification by enabling the free exchange of ideas and, on the other hand, alienating the participants?

   Interpretation: ...isn't it strange to see that, on the one hand, the Internet _____ and, on the other hand, makes it difficult for the participants to feel comfortable with one another?

3. **Text sentence:** I had a recent discussion with a friend of mine who has what he described as a problem with the Internet.

   Interpretation: I discussed _____, and he said that he had a problem with the Internet.

4. **Text sentence:** ... all of the time that he spent online might have altered his sense of reality, and that it made him feel lonely and depressed.

   Interpretation: ... because _____ sense of reality, and as a result felt lonely and depressed.

5. **Text sentence:** It seems to me that the Internet is a powerful tool that presents an opportunity for the advancement of the acquisition and application of knowledge.

   Interpretation: It seems to me that the Internet is a powerful tool, because it _____.

6. **Text sentence:** However, based on my personal experience I can understand how, as they surf the Web, some folks might be confronted with cognitive overload.

   Interpretation: My personal experience helps me to understand that _____, they might be confronted with cognitive overload.

7. **Text sentence:** Depending upon how a real place is defined it might very well be.

   Interpretation: The Internet could possibly be considered as a real place _____.

8. **Text sentence:** Some would argue that the virtual existences created by some users who debate, shop, travel and have romance online are in fact not real.

   Interpretation: Some people do not think the Internet is a real place, because _____ _____.

9. **Text sentence:** While others would argue that, since in practical terms, folks are debating, shopping, traveling and having romance, the opposite is true.

   Interpretation: Other people would say that the Internet is a real place, because computer users are actually _____.

C. Detailed Study of the Text

----

1 **However, much less widely reported has been the idea that the Internet may be responsible for furthering the splitting up of society by alienating its individual users.** (Para. 1) 但鲜有报道提到使用互联网可能导致使用者相互疏离，从而加速整个社会的分裂。

   much less widely reported has been the idea... 是一个谓语前置的倒装句，正常的语序应该是 the idea... has been much less widely reported, idea 后面的从句为同位语从句。

2 **My brother-in-law and I were chatting nearby and it struck us that in generations past, his son, my nephew, would have been outside playing with his friends.** (Para. 3) 我的姐夫和我在一旁聊天，这使我突然想到若是在以前，他的儿子，即我的外甥，应该和他的朋友在室外玩耍。

   句中的 strike 意为"闯入脑海，想到"。如：The thought struck me all of a sudden. 我突然有了这个想法。

3 Depending upon how a "real place" is defined it might very well be. (Para. 5) 根据"真实的地方"的定义，那么互联网也完全可能是真实的。

这里的 it might very well be 说完整是 it might very well be a real place。

4 While others would argue that, since in practical terms, folks are debating, shopping, traveling and having romance, the opposite is true. (Para. 5) 其他人会争辩说，事实上，网民们在网上讨论、购物、旅游和恋爱，因此虚拟世界是真实的。

the opposite is true 是指情况与前文所说的虚拟世界是假的恰恰相反。

D. Talking About the Text

Work in pairs. Ask and answer the following questions first and then put your answers together to make an oral composition.

1. Do you agree with the idea that the Internet may be responsible for furthering the splitting up of society by alienating its individual users?
2. What problem did the author's friend have with the Internet?
3. Why did the author's friend say that he felt lonely and depressed?
4. Give an example to show that the younger generation prefers to play with friends in cyberspace.
5. What problems will some people probably meet with when they surf the Web?
6. What do you think of the Internet? Is it a real place or not?
7. Do you think the information superhighway will lead us to utopia? Why or why not?

E. Vocabulary Practice

I. Fill in the blanks with the new words or expressions from Text A.

1. The _____ gap can be overcome through effective communication.
2. I find no _____ between his publicly expressed opinions and his private actions.
3. When people get _____ to the Internet, it will be difficult for them to go off-line.
4. The knowledge you gained will change the way you _____ the world.
5. Hard work is _____ to success.
6. They organized a dinner in _____ of the year's successes.
7. For the purposes of the survey we've _____ the town into four areas.
8. The government has _____ the ban on the import of beef until June.

9. The book gave a _____ picture of his childhood.

10. He devotes his time to the _____ of knowledge.

II. Complete the following dialogs with appropriate words or expressions from Text A.

1. A: What is the characteristic of the _____ age?

   B: It is marked by the wide use of computers and adoption of the Internet.

2. A: Can you name one disadvantage of the Internet?

   B: Yes, it can _____ its users from other people.

3. A: He didn't mean to hurt them.

   B: But the joke he told brought about _____ consequences.

4. A: You are unsatisfied with these young people.

   B: Yes, I think the younger _____ are less considerate and they are too noisy.

5. A: Let's have a virtual tour of this famous university.

   B: OK, but later I will have a _____ investigation this coming summer.

6. A: She looks very happy as she can finally go home now.

   B: Yes, but when she was told that the plane was delayed again, she became really _____.

7. A: Why is this discovery so important to the local people?

   B: Because it _____ the history of the local civilization.

8. A: I'm so tired.

   B: How about a cup of coffee at a _____ coffee shop?

9. A: I don't like this romance story.

   B: I agree with you. It doesn't have any sense of _____.

10. A: The child spent a lot of his time reading.

    B: Reading helps to develop one's _____ ability.

 Before Reading

Discuss the following questions in class.

1. If one of your best friends is going to break up with you, what would you do and how would you feel?

2. Do you think that some of the problems in your personal relationships can be solved by yourself?

 **Reading**

Dealing with Problems in Your Personal Relationships

1   Volumes of books and extensive courses have been created to explore the infinite[1] complexity[2] of human relationships. Problems can arise[3] from a large number of sources and it can frequently need some care to help sort out[4] the influences. These problems can become more serious because of the pressures from others to form or end a relationship and the general pressures from the media which give an idealised view of couples which is often at odds[5] with the reality many people experience.

2   Here are some simple guidelines to help you explore and reduce tensions[6] which you may be feeling about relationships.

3   1. Do you know what you are looking for in a relationship? There are many different reasons for entering into a relationship—for companionship[7], to have a long-term partner, to create a family and so on. Do you know what you are looking for? Have you discussed this with your partner? If not, there is a distinct[8] possibility that you may both end up seriously at cross-purposes.

4   2. Are you asking too much or expecting too little from your relationship? A good relationship can provide support, companionship and eventually an opportunity to build a joint[9] life. If you are looking to it to provide more than this—for example, to give you a sense of purpose and worth or protect you from some deep personal fear—you may be trying to get a partner to provide things that in fact only you can achieve. If on the other hand a relationship brings you continual[10] sorrow and unhappiness you may be accepting for yourself a far lower level of interaction than you have a right to expect. In particular no one deserves[11] to be on the receiving end of physical violence. Do look for the support you need to change or end a relationship if violence is happening to you.

5   3. Have you got a model for the relationship you are trying to build? Many people find it helpful to picture a relationship that they admire and into which they wish to enter. It may be the relationship of someone you know or an imagined one. Consider how the people in this relationship settle[12] differences and overcome difficulties. If it is not obvious and the relationship is a real one, ask them. If they have never been seen to have any problems, maybe they are not a terribly realistic model after all!

6   4. Can you talk about problems? In all relationships there are going to be times of serious disagreement[13], where a conflict of interests has to be settled. This doesn't

mean there is something wrong with the relationship. However, arguing the point out and reaching agreement does take a bit of skill and practice. Many relationship advisers suggest the best way to solve a relationship problem is to speak for up to fifteen minutes about your view of the problem. The other person listens carefully, interrupting[14] only to clarify[15] and to help you express yourself clearly. Then the other person takes a similar time to explain their point of view. Finally take half an hour to talk together to see if you can mend[16] the difference. If you don't succeed this time, return to the problem a few days later and try again.

7   If you are not in the habit of talking in your relationship, it might be interesting to give it a try. Relationships are one of the curious features of human existence and can be well worth exploring.

New Words and Expressions

1 infinite ★ /'ɪnfɪnət/ a. 无限的，无穷的，不确定的

2 complexity /kəm'pleksətɪ/ n. 复杂性

3 arise /ə'raɪz/ v. （由……）引起，起源（于）

4 sort out 澄清

5 at odds (with) 与……不一致

6 tension ▲ /'tenʃən/ n. 紧张，紧张状态

7 companionship /kəm'pænjənʃɪp/ n. 友谊，交情

8 distinct ★ /dɪs'tɪŋkt/ a. 明确的，显著的

9 joint ★ /dʒɔɪnt/ a. 共同的，共有的

10 continual ★ /kən'tɪnjʊəl/ a. 不间断的，不停的

11 deserve ★ /dɪ'zɜːv/ v. 应受，应得

12 settle ★ /'setl/ v. 解决

13 disagreement # /ˌdɪsə'griːmənt/ n. 意见不同，争执
     agreement /ə'griːmənt/ n. 一致，感情融洽

14 interrupt /ˌɪntə'rʌpt/ v. 打断

15 clarify ★ /'klærɪfaɪ/ vt. 澄清，阐明

16 mend /mend/ vt. 修理，修补

After Reading

A. Main Idea

Complete the following diagram with the sentences or expressions given below.

1. for companionship, partnership or friendship
2. How much do you expect from a relationship?
3. What kind of model should you learn from?
4. How to settle disputes in a relationship?
5. Do you know your purpose of entering into a relationship?
6. You may expect a relationship to provide support, companionship and eventually an opportunity to build a joint life.
7. An appropriate model contributes to a successful relationship.
8. to have a bit of skill and practice

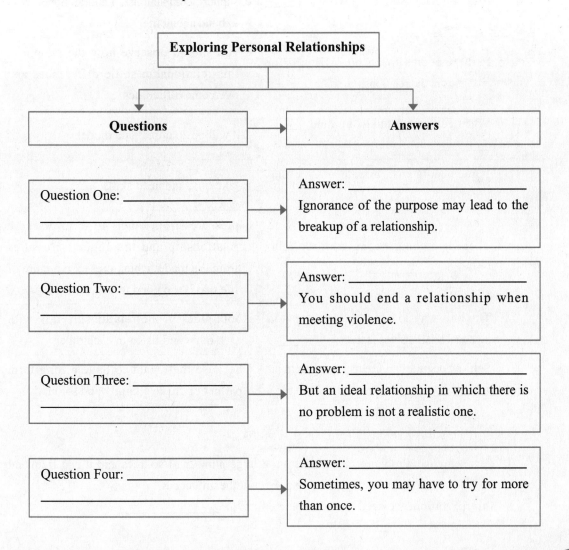

B. Detailed Understanding

I. Make correct statements according to the text by combining appropriate sentence parts in Column A with those in Column B.

Column A	Column B
1. A large number of books and lectures discuss _____.	a. when the people concerned have disputes or experience differences
2. Problems in personal relationships may arise from _____.	b. they may have different purposes for the relationship and thus end in separation
3. If two partners do not communicate enough with each other, _____.	c. how complex human relationships are
4. A good relationship can give you _____ _____.	d. support, companionship and an opportunity to build a joint life
5. If there seems to be no difficulty or problems in the relationship, _____.	e. you should observe how the people in this relationship settle differences and overcome difficulties
6. When you have a model in mind, _____ _____.	f. it will be an unrealistic model
7. Some people wish to enter into a relationship _____.	g. take time and need skills
8. There is nothing wrong with the relationship _____.	h. pressures from others to form or end a relationship and the general pressures from the media which present a picture of life that is too good to be true
9. The best way to solve the problems in personal relationships is that _____.	i. both sides in the relationship talk with each other and listen to each other
10. Settling disputes and reaching agreement _____.	j. because they want to have a long-term partner or create a family and so on

II. Explain the sentences by filling in the blanks.

1. **Text sentence:** Problems can arise from a large number of sources and it can frequently need some care to help sort out the influences.

   Interpretation: We need to make some efforts to _____.

2. **Text sentence:** ... the general pressures from the media which give an idealised view of couples which is often at odds with the reality many people experience.

   Interpretation: .... the mass media often describe a perfect picture of the relationship which will make the ordinary people feel pressured because _____ _____.

3. **Text sentence:** ... there is a distinct possibility that you may both end up seriously at cross-purposes.

   Interpretation: ... it is quite possible that you and your partner may _____.

4. **Text sentence:** ... you may be trying to get a partner to provide things that in fact only you can achieve.

   Interpretation: ... maybe the things you expect your partner to give _____.

5. **Text sentence:** ...you may be accepting for yourself a far lower level of interaction than you have a right to expect.

   Interpretation: ... you have the right to expect _____.

6. **Text sentence:** In particular no one deserves to be on the receiving end of physical violence.

   Interpretation: What needs to be noted particularly is that _____.

7. **Text sentence:** Many people find it helpful to picture a relationship that they admire and into which they wish to enter.

   Interpretation: Before entering into a new relationship, _____.

8. **Text sentence:** If they have never been seen to have any problems, maybe they are not a terribly realistic model after all!

   Interpretation: If no problem has been seen in a certain relationship, _____!

9. **Text sentence:** The other person listens carefully, interrupting only to clarify and to help you express yourself clearly.

   Interpretation: The other side should listen to you carefully and only stop you to _____.

C. Detailed Study of the Text

1 **Volumes of books and extensive courses have been created to explore the infinite complexity of human relationships.** (Para. 1) 人们写了大量的书，开设了无数的课程来研究人际关系的复杂性。

human relationship 意为"人际关系"。其他类似的短语如：human resources "人力资源"，human rights "人权"等。

2 **...the media which give an idealized view of couples which is often at odds with the reality many people experience.** (Para. 1) 媒体所展示的理想化的夫妻关系与现实生活中许多人的经历不吻合。

at odds (with) 等于 have quite different opinions，意为"争吵不合，有分歧"。如：It is a strange family and all the family members are at odds with one another. 这是一个奇怪的家庭，每个家庭成员都和其他成员不合。另外，which is often at odds with... 为定语从句，修饰 an idealized view of couples。

3 **There are many different reasons for entering into a relationship—for companionship, to have a long-term partner, to create a family and so on.** (Para. 3) 人们因为不同的原因而建立某种关系：有的为了交友，有的为了有一个长期的同伴，有的为了建立家庭，等等。

partner 有"合伙人，伙伴"的意思，如：Peter and Mary are partners in a law firm. 彼得和玛丽合伙开了一家律师事务所。

4 **In particular no one deserves to be on the receiving end of physical violence.** (Para. 4) 尤其是，没有人应该成为暴力行为的受害者。

be on the receiving end 意为"成为接受方"。physical violence 指伤害身体的暴力行为。

5 **The other person listens carefully, interrupting only to clarify and to help you express yourself clearly.** (Para. 6) 另一方仔细地倾听，偶尔打断也只是为了澄清观点或是帮你更清楚地表达。

only to 引导一个目的状语，等同于 in order to, so as to, as to 等。如：He ran so fast as to catch the first bus. 他飞快地跑以便赶上第一班车。I come here only to say goodbye to you. 我来就为了向你告别。

D. Further Work on the Text

Write down at least three more comprehension questions of your own. Work in pairs and ask each other these questions. If you can't answer any of these questions, ask your classmates or the teacher for help.

1. _____

2. _____

3. _____

E. Vocabulary Practice

I. **Find the word that does NOT belong to each group.**

1. A. distinct   B. eventually   C. personal   D. continual
2. A. pressure   B. stress   C. motivation   D. burden
3. A. couple   B. people   C. companion   D. problem
4. A. relationship   B. friendship   C. partnership   D. companionship
5. A. state   B. indicate   C. point out   D. figure out

II. **Complete the following sentences with appropriate words in their correct form.**

1. **distinct, distinctive, distinction**
   1) Alcohol has a very _____ smell; it's quite _____ from the smell of wine.
   2) There is a _____ possibility that she won't come.
   3) A _____ should be made between the two issues.

2. **explore, exploration**
   1) They have _____ all possibilities of reaching an agreement.
   2) Tony has begun his _____ to outer space.

3. **companion, companionship**
   1) He was my only Chinese _____ during my stay in Australia.
   2) Mary valued the _____ between her and her husband.

4. **create, creative, creation**
   1) Criticizing will only destroy a relationship and _____ feelings of failure.
   2) Language is the most important _____ of man.
   3) This job is so boring. I wish I could do something more _____.

5. **realistic, realism, realize, realization, real, reality**
   1) The price of petrol goes up so quickly, so we have to be _____ and give up our car.
   2) All science is basically searching for _____.
   3) The _____ of his lifelong dream makes him break into tears.
   4) He has made a big mistake but he didn't _____ it.
   5) There is much _____ in what you say and I really appreciate your down-to-earth attitude.
   6) He is a _____ man and he always helps the people in need.

Word Study

further

ad. 1. 更大程度上，进一步地：She should consider further the consequences of her actions. 她应该更多地考虑她的行为所造成的后果。

2. 而且；此外：He stated further that he would not cooperate with the committee. 他还表明他不愿与委员会合作。

3. 更远；再往前地：I went only three miles further yesterday. 昨天我才往前走了 3 英里。

a. 1. 更多的；另外的：I have nothing further to say. 我没有别的话要说了。

2. 更远的：It is a result that is further from our expectations than last time. 这是一个比上一次更出乎我们意料的结果。

v. 促进；推进：Your visit will further the relationship between our two parties. 你的来访将促进我们双方的关系。

cross

v. 1. 穿过，越过：He crossed the room to greet us. 他穿过房间来迎接我们。

2. （使）交叉，（使）相交：Elm Street crosses Oak Street. 榆树街和橡树街相交叉。

n. 十字形：He made the sign of the cross upon his chest as a sign of devotion. 他在胸口划十字以示虔诚。

opposite

prep. 在……对面：He parked the car opposite the bank. 他将车停在银行的对面。

ad. 在对面：They sat opposite at the table. 他们面对面地坐在桌子两旁。

a. 1. 对面的：The library is on the opposite side of the road from the school. 图书馆在学校马路对面。

2. 相反的，对立的：The effect of the medication was opposite to that intended. 这种药物的反应和预期的相反。

n. 对立面，对立物：High is the opposite of low. 高是低的对立面。

派 opposition

strike

v. 1. 打，击，撞：She struck the desk with her knee. 她的膝盖撞上了书桌。

2. 罢工：The workers were striking because they wanted higher pay. 工人们在罢工，因为他们要求增加工资。

3. 给……以印象，使受吸引：How does the idea strike you? 你觉得那主意怎么样？

4. （钟等）敲响；报点：The clock struck nine. 钟敲了 9 点。

5. （使）突然想起；（使）认为：An idea suddenly struck me. 我突然有了这个主意。

n. 罢工：The strike is due to begin on Tuesday. 罢工定于星期二开始。

on strike 罢工：Most of the employees were on strike. 许多雇员在罢工。

influence

n. 1. 影响：The tribe remains untouched by outside influences. 这个部落仍然未受到外界的影响。

2. 势力；权势：She used her parents' influence to get the job. 她利用父母的权势找到了工作。

v. 影响；对……起作用：My teacher influenced my decision to study science. 我的老师对我学理科的决定是有影响的。

Writing Practice

I. Fill in the blanks with the proper words given below.

> bottle   can   loaf   crate   carton   box   packet   shower   article   flash   piece

1. two _____ of milk
2. a _____ of skim milk
3. ten _____ of Coke
4. a _____ of cookies
5. a _____ of bread
6. two _____ of beer
7. a _____ of soy sauce
8. four _____ of juice
9. six _____ of tissue
10. a _____ of rain
11. an _____ of clothing
12. a _____ of lightning
13. a _____ of news

II. Add a measure word before the underlined noun.

1. My mother gave me <u>advice</u> that I have always remembered.
2. I can give you important <u>information</u> about that.
3. There is new <u>equipment</u> in that area.
4. I need to get some <u>furniture</u> for my apartment.
5. Beethoven wrote wonderful <u>music</u>.
6. How much <u>luggage</u> can I check in?

III. Rewrite the sentences according to the models.

Model A:

Original sentence: I had a recent discussion with a friend of mine who has something that he described as a "problem" with the Internet. (Text A)

New sentence: I had a recent discussion with a friend of mine who has what he described as a "problem" with the Internet.

1. I had a recent talk with my uncle who has something that he described as a "problem" with his work.

   _____

   _____

2. Peter had a recent discussion with Mary who has something that she described as a "problem" with alcohol.

   _____

   _____

Model B:

Original sentence: The Internet may result in furthering the splitting up of society by alienating its individual users. (Text A)

New sentence: The Internet may be responsible for furthering the splitting up of society by alienating its individual users.

3. According to the report, it was a problem with the engine that resulted in the crash.

   _____

   _____

4. The bus driver's carelessness resulted in the accident.

   _____

   _____

Model C:

Original sentence: My brother-in-law and I were chatting nearby and it occurred to us that in generations past, his son, my nephew, would have been outside playing with his friends. (Text A)

New sentence: My brother-in-law and I were chatting nearby and it struck us that in generations past, his son, my nephew, would have been outside playing with his friends.

5. I was helping my daughter with her homework and it occurred to me that in my childhood I hardly worried about homework after school.

    _____

    _____

6. I was reading a novel and it occurred to me that the heroine and I had much in common.

    _____

    _____

Model D:

Original sentence: In particular no one should be on the receiving end of physical violence. (Text B)

New sentence: In particular no one deserves to be on the receiving end of physical violence.

7. In particular, your friend Peter should be rewarded.

    _____

    _____

8. This research paper should receive your attention.

    _____

    _____

IV. **Combine each set of the sentences into one, using the connective words and expressions provided.**

1. a. He said he spent a lot of the time online.
    b. This might have altered his sense of reality.
    c. This might have made him feel lonely and depressed.
    New sentence: _____ (that, and thus)
        (Text A)

2. a. During the celebration my young nephew spent the entire time on my computer.
    b. He was playing a virtual war game.
    New sentence: _____ (-ing)
        (Text A)

3.   a. Some people have a belief.

    b. "Information superhighway" will lead us to utopia.

    c. Some other people feel it is the road to hell.

    New sentence: _____ (while)

        (Text A)

4.   a. At the very least, I believe something.

    b. We use the Internet.

    c. We are forced to ask fundamental questions about something.

    d. We perceive the world about us.

    New sentence: _____ (that, when, how)

        (Text B)

5.   a. These problems can become more serious.

    b. There pressures from others can form or end a relationship.

    c. The general pressures from the media give an idealised view of couples.

    d. This idealized view of couples is often at odds with the reality many people experience.

    New sentence: _____

    (because of, and, which, which) (Text B)

6.   a. A relationship brings you continual sorrow and unhappiness.

    b. You may be accepting for yourself a far lower level of interaction.

    c. You have a right to expect something.

    New sentence: _____ (if, than)

        (Text B)

7.   a. Many people find it helpful to picture a relationship.

    b. They admire a relationship.

    c. They wish to enter into a relationship.

    New sentence: _____ (that, and, which)

        (Text B)

8.   a. In all relationships there are going to be times of serious disagreement.

    b. There a conflict of interests has to be settled.

    New sentence: _____ (where)

        (Text B)

V. Translate the following sentences into English.

1. 公共汽车司机应对乘客的安全负责。（be responsible for）

2. 电视一方面丰富了人们的业余生活，另一方面却减少了家人交流的时间。（on the one hand, on the other hand）

3. 她突然想到她可能再也回不来了。（strike）

4. 她在警察让她面对证据时才承认偷了钱。 （confront with）

5. 她不很喜欢他；事实上，我认为她恨他！ （in fact）

6. 玛丽总是提醒彼得，祸从口出。 （arise from）

7. 大家都认为张红的行为应受到表彰。 （deserve）

8. 我不清楚他们俩是如何消除差异并克服困难的。 （settle, overcome）

VI. Translate the following paragraph into Chinese.

Married men and women sometimes consider each other best friends as well as spouses. They also socialize with members of the opposite sex either as couples or independently. A working wife may have a close male friend at her job. This does not mean that there are no constraints on the married men or women. The limitations of these relationships are not always visible but they do exist. Most, but not all, married American couples practice monogamous (一夫一妻的) relationships.

VII. Practical English Writing

Directions: The Internet has been with us for many years. Some say that the Internet has brought people closer. For example, with the help of the Internet, you can now send e-mails to any person anywhere at any time at little or even no cost. But others say that the Internet has pulled people apart from each other. People rely so much on the Internet that they no longer see each other or talk to each other. Now discuss with your classmates and see if you can find more examples that show the influence of the Internet on interpersonal relationships.

After discussion, please write a passage on the following topic:

Influence of the Internet on Interpersonal Relationships

4
Unit

Before Reading

Discuss the following questions in class.

1. When you hear the word culture, what do you associate it with? Do you think that a company can also have its own culture?

2. What do you think are the most important things for a successful company?

Reading

○   ○

Corporate¹ Culture: A Case Study

1   CORPORATE CULTURE is loosely² defined as the attitudes, behaviors and personalities that make up a company. In other words, it is how we view our work and ourselves. If we accept this general definition, the next thought is: How does it apply? Through my consulting³, articles, website and radio show, I have been asked the question, Yeah⁴, we know what it is—but what does it do? Fortunately⁵, and unfortunately, I have been an eyewitness⁶ to a fascinating⁷ case study. My case study involved two similar businesses, about the same size, and in the same industry. Both were struggling financially⁸.

2   Upon my initial⁹ analyses, both businesses had good potential and both retained¹⁰ me to help them grow, create wealth and sustain profitability¹¹. Both had very similar problems and both had owners that were ego¹²-driven and hard workers. There was never a question in either company of the willingness to work hard. There was, however, a great deal of difference in the results.

3   After my analyses and employee interviews, I determined that both owners were holding their businesses back. Both owners acknowledged they were a problem in their own companies. The owner of Company A became convinced¹³ he was such a problem that, for his business to grow, he had to leave it. He turned his decision-making and management over to me. The owner of Company B also acknowledged he was part of the problem, but decided that by working harder, he could overcome the problems he created.

4   The first thing I did at Company A was to fire some minor employees and hire

some better ones. I then turned the company over to them. The absentee[14] owner of Company A expressed his concern at doing this but accepted it. He understood there was no alternative[15]. I walked the managers through some tough decisions and encouraged them. They made mistakes but I made certain the mistakes were small ones and I encouraged them to learn and move on.

5   After several months, some very interesting developments occurred[16]: (a) a fierce company loyalty developed among all employees; (b) they would not let the absentee owner make any decisions; (c) my intervention[17] became less and less necessary—all employees constantly discussed how to improve productivity[18] and deliver more value to the customer; (d) profitability increased to the point that all employees got raises; (e) morale[19] steadily[20] improved; (f) Company A began to gain market share.

6   Company B took a different route[21]. The owner did not want to fire any minor employees because he had become a friend and "father-figure" to them. The owner began to work longer and longer hours. He began to distrust his best people. After several months, some interesting developments occurred: (a) the stress level of all employees went up; (b) several key people quit; (c) Company B was not able to attract good employees; (d) employees began to feel unhappy about the management style and looked for ways to get back at[22] the company; (e) more and more intervention was necessary on my part to keep the status quo[23]; (f) profitability decreased[24] and customers were lost.

7   Six months later, the results were not surprising. Company A was growing steadily, morale was high and their employees were the highest paid in the industry. Employees enjoyed coming to work and worked very hard. They constantly were looking for ways to improve and look for new customers and markets.

8   Company B downsized[25] and filed for protected bankruptcy[26]. Employees were discouraged and many began looking elsewhere[27] for work. Customers noticed that Company B was in trouble and took their business elsewhere.

9   These two examples are extremes[28] and I was most fortunate at having the opportunity to carefully examine both. I think about them both quite often and have resolved[29] to make corporate culture an even higher priority in my work.

10   Since people drive a business, corporate culture has become the vehicle to get to the desired destination[30].

New Words and Expressions

1   corporate■ /'kɔːpərət/ *a.* （法人）团体的，公司的

2   loosely# /'luːslɪ/ *ad.* 不严谨地，松散地

3   consult /kən'sʌlt/ *v.* 请教，查阅

4   yeah /jeə/ *ad.* （口）= yes ［变体］

5   fortunately /'fɔːtʃənətlɪ/ *ad.* 幸运地，幸亏

6   eyewitness# /'aɪˌwɪtnɪs/ *n.* 目击者；见证人

7   fascinating■ /'fæsɪneɪtɪŋ/ *a.* 有极大吸引力的

8   financially# /faɪ'nænʃəlɪ/ *ad.* （在）财政方面，（在）金融方面

9   initial★ /ɪ'nɪʃəl/ *a.* 开始的，最初的

10 retain★ /rɪ'teɪn/ *v.* 付定金聘请（尤指律师、顾问等）

11 profitability# /ˌprɒfɪtə'bɪlətɪ/ *n.* 利润；利益

12 owner /'əʊnə(r)/ *n.* 物主，所有人

13 ego■ /'iːgəʊ/ *n.* 自我，自己

14 absentee# /ˌæbsən'tiː/ *n.* 不在者，缺席者

15 alternative /ɔːl'tɜːnətɪv/ *n.* 供替代的选择

16 occur /ə'kɜː(r)/ *v.* 发生，出现

17 intervention■ /ˌɪntə'venʃn/ *n.* 介入，干涉，干预

18 productivity■ /ˌprɒdʌk'tɪvətɪ/ *n.* 生产力，生产率

19 morale■ /mɒ'rɑːl/ *n.* 士气，斗志

20 steadily# /'stedɪlɪ/ *ad.* 稳定地

21 route★ /ruːt/ *n.* 路线，路程

22 get back at 报复

23 status quo *n.* 现状

24 decrease /dɪ'kriːs/ *v.* 减小，减少

25 downsize# /'daʊnsaɪz/ *v.* 裁员

26 bankruptcy /'bæŋkrəptsɪ/ *n.* 破产

27 elsewhere /ˌels'hweə(r)/ *ad.* 在别处，到别处

28 extreme /ɪk'striːm/ *n.* 极端

29 resolve▲ /rɪ'zɒlv/ *vt.* 解决，解答

30 destination★ /ˌdestɪ'neɪʃən/ *n.* 目的地，终点

After Reading

A. Main Idea

Complete the following diagram with the sentences or expressions given below.

1. beginning to recover and develop steadily
2. Corporate culture is the key to success in a company.
3. changing its corporate culture
4. refusing to change its corporate culture
5. Both companies had owners that were ego-driven and hard workers. Both owners were holding back the development of the companies.
6. how the people of a company view their work and themselves
7. going downhill to the point of bankruptcy

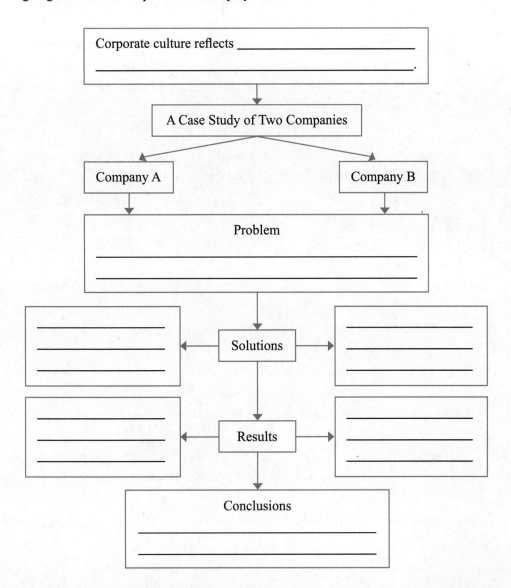

B. Detailed Understanding

I. Tell if the following statements are true (T) or false (F) according to the text.

1. _____ Corporate culture can be understood as the attitudes, behaviors and personalities that represent a company.

2. _____ The author's case study was about two entirely different companies.

3. _____ The owner of Company A didn't realize he was the problem in his company.

4. _____ According to the author's initial analyses, both companies had the possibility to become successful.

5. _____ After several months, the employees of Company A became more loyal to and more confident in the company.

6. _____ After several months, Company B began to gain profit.

7. _____ The author had to be more involved in the management of Company B in order to maintain the situation as it was.

8. _____ The case study of the two companies shows how important corporate culture is and how it works.

9. _____ According to the author, corporate culture is like a car you can drive to the place you want to go.

10. _____ People can easily understand what corporate culture is and how it functions.

II. Explain the sentences by filling in the blanks.

1. **Text sentence:** Upon my initial analyses, both businesses had good potential and both retained me to help them grow, create wealth and sustain profitability.

   Interpretation: According to my initial analyses, both companies _____.

2. **Text sentence:** There was never a question in either company of the willingness to work hard.

   Interpretation: There was no doubt that _____.

3. **Text sentence:** The owner of Company A became convinced he was such a problem that, for his business to grow, he had to leave it.

   Interpretation: The owner of Company A became aware that _____.

4. **Text sentence:** The owner of Company B also acknowledged he was part of the problem, but decided that by working harder, he could overcome the problems he created.

   Interpretation: The owner of Company B also admitted that _____.

5. **Text sentence:** I walked the managers through some tough decisions and encouraged them.
   Interpretation: I helped the managers make _____.

6. **Text sentence:** ... my intervention became less and less necessary—all employees constantly discussed how to improve productivity and deliver more value to the customer...
   Interpretation: ... I was less and less involved in _____...

7. **Text sentence:** Employees began to feel unhappy about the management style and looked for ways to get back at the company.
   Interpretation: Employees began to feel unhappy about _____.

8. **Text sentence:** I think about them both quite often and have resolved to make corporate culture an even higher priority in my work.
   Interpretation: I often think about these two companies and _____.

9. **Text sentence:** Since people drive a business, corporate culture has become the vehicle to get to the desired destination.
   Interpretation: Since people run a business just like driving a vehicle, _____.

C. Detailed Study of the Text

1　**The owner of Company A became convinced he was such a problem that, for his business to grow, he had to leave it.** (Para. 3) A 公司的老板确信自己是问题所在，因此为了公司的发展，他必须离开。
句中 such... that 的意思为"如此……以至于"，such 后跟名词。

2　**The owner of Company B also acknowledged he was part of the problem, but decided that by working harder, he could overcome the problems he created.** (Para. 3) B 公司的老板也承认他是问题之一，但却认定他可以通过更努力地工作来解决他惹出的问题。
句中 decided 后跟由 that 引导的宾语从句。he created 为 problems 的定语从句。

3　**I walked the managers through some tough decisions and encouraged them.** (Para. 4) 我带领公司的管理者们一起做出了一些痛苦的抉择，并激励他们。
句中 walk somebody through something 的意思为"带某人一起从事某事"。

4　**protected bankruptcy** (Para. 8) 破产保护
美国《破产法》管辖着公司如何停止经营或如何走出债务深渊的行为。当一个公司临近山穷水尽之境地时，可以援引《破产法》第十一章来"重组"业务，争取再度赢利。如果依据《破产法》第七章申请破产，公司全部业务必须立即完全停止。多数上市公司会按照《破产法》第十一章申请破产保护，而不是按照第七章，直接进行破产清算，因为他们仍希望继续运营并控制破产程序。

5 **Since people drive a business, corporate culture has become the vehicle to get to the desired destination.** (Para. 10) 因为经营靠的是人，企业文化则如同车辆一样可以帮助人们到达所期望的目的地。

此句是暗喻。将人们经营企业比作开车。因此，corporate culture 就被形象地描述成 vehicle，人们可以用此 vehicle 到达自己想去的目的地。

D. Talking About the Text

Work in pairs. Ask and answer the following questions first and then put your answers together to make an oral composition.

1. What is corporate culture?
2. How does the author explain the importance of corporate culture?
3. What did the author know upon his initial analyses?
4. What was the common problem for these two companies?
5. What did the owner of Company A do?
6. What did the owner of Company B do?
7. What happened to Company A?
8. What happened to Company B?
9. What do you think of corporate culture after reading this article?

E. Vocabulary Practice

I. Fill in the blanks with the new words or expressions from Text A.

1. The _____ of the enemy troops is low.
2. You can _____ the right to reply.
3. There was a _____ fight between the two men.
4. The two men would have continued fighting but for the _____ of the policeman.
5. We should be focusing on _____ kids from smoking.
6. Managers want to maintain the _____ because they're afraid to take any risks.
7. If a company _____, it reduces the number of people it employs in order to reduce costs.

8. We weren't sure about which _____ we should take.

9. There weren't enough beds, but the matter was _____ by Tom sleeping on the sofa.

10. It never _____ to me for a moment that she meant she was quitting the job.

II. **Complete the following dialogs with appropriate words or expressions from Text A.**

1. A: Our dog keeps barking all the time. Our neighbors have complained many times.
   B: We have to try and change his _____.

2. A: We're going on our summer vacation tomorrow.
   B: _____, the weatherman predicts nice sunny weather.

3. A: The Museum of History's special exhibition opens this weekend.
   B: It will be _____ to see what the latest discoveries are.

4. A: The journalists worked hard on reporting the tragic events.
   B: Their _____ of the situation was accurate.

5. A: There are too many students applying for the elective course I intended to take.
   B: You may have to find some _____.

6. A: It was not the first time that she had heard this story. Her best friend had lied to her.
   B: Now her _____ of her friend was complete.

7. A: We are traveling to Europe this summer.
   B: Our main _____ will be France.

8. A: Even with technology and satellites, weather forecasting is still not accurate.
   B: While the weather patterns in the west continue to be predictable, _____, this has not been the case.

9. A: The university keeps growing larger and larger.
   B: They are _____ increasing the number of classes they offer.

10. A: The company had to close several of its factories because of the failing economy.
    B: Unfortunately, this meant it had to _____ the number of its employees too.

Text B

 Before Reading

Discuss the following questions in class.

1. If you own a company, what kind of corporate culture would you like to build? Why?
2. What do you think is a healthy corporate culture?

 Reading

Healthy Corporate Cultures

1   Generally, corporate culture refers to the prevailing[1] unspoken values, attitudes and ways of doing things in a company. It often reflects the personality, philosophy and the cultural background of the founder or the leader. Corporate culture affects how the company is run and how people are promoted.

2   Positive corporate cultures create a positive work climate, which can lead to productivity and job satisfaction. They contribute[2] to high performance without apparently linking reward to performance. The ideal company should possess the characteristics[3] of the following four types of healthy corporate cultures.

3   1. Progressive[4]-adaptive[5] culture—There is openness to new ideas and a willingness to take risks and adopt innovations[6]. It is a culture that adjusts quickly to shifting market conditions. It does not value the certainty of remaining the same; the only certainty it values is that the company is future-oriented[7] and innovative. It is confident in catching and riding the waves of change.

4   It is a culture compatible[8] with the spirit of creativity. The management makes every effort to be on the cutting edge, and encourages continuous development of workers. There is a restless creative energy, constantly seeking and creating new ideas and new markets. The company celebrates every innovation, and every discovery. Excitement is in the air. Employees are all caught up in the adventure[9].

5   2. Purpose-driven culture—The leadership communicates the purpose of the company effectively, so that there is a common purpose, a shared vision for all the workers. Everyone knows what the core[10] values and priorities[11] are, and everyone knows where the company is going. Workers are highly motivated, because they are committed to the same set of core values.

6   3. Community-oriented culture—There is a strong emphasis on cooperation[12]. The

leadership attempts to build a community, in which people respect and support each other, and enjoy working together. A community-oriented culture goes beyond team building and seeks to create a genuine[13] community in which every worker is treated as a valuable member. Community building is more extensive than team building. It requires that members from different work groups treat each other in a positive, supportive way in order to enhance[14] morale.

7   For teamwork to be effective, team building training becomes an important part of personnel development. Typically in team building, groups are created in each work area and group members interact[15] and work together to identify and resolve issues that affect individual and group performance. Guidelines are provided for interactions among team members.

8   4. People-centred culture—There is a genuine caring for each worker in the organization. Everyone is valued, regardless of[16] their positions in the company. The organization cares for the whole person in terms of recognizing workers' basic needs for learning and growth, for belonging and being connected. Each worker is encouraged to develop his or her full potentials, personally and professionally. Such a culture will create a climate of mutual[17] respect.

New Words and Expressions

1 prevailing /prɪ'veɪlɪŋ/ *a.* 流行的，盛行的
   prevail＊ /prɪ'veɪl/ *v.* 流行，盛行
2 contribute /kən'trɪbjuːt/ *v.* 起促成作用
3 characteristic /ˌkærəktə'rɪstɪk/ *n.* 特性，特征
4 progressive /prəʊ'gresɪv/ *a.* 进步的，先进的
5 adaptive /ə'dæptɪv/ *a.* 适应的，有适应性的
6 innovation /ˌɪnəʊ'veɪʃən/ *n.* 创新
   innovative /'ɪnəʊveɪtɪv/ *a.* 富有革新精神的
7 orient＊ /'ɔːrɪənt/ *vt.* 使朝向，以……为方向
8 compatible /kəm'pætəbl/ *a.* 兼容的
9 adventure /əd'ventʃə(r)/ *n.* 异乎寻常的经历
10 core /kɔː(r)/ *a.* 核心的，主要的
11 priority＊ /praɪ'ɒrətɪ/ *n.* 优先考虑的事
12 cooperation /kəʊˌɒpə'reɪʃən/ *n.* 合作，配合
13 genuine /'dʒenjʊɪn/ *a.* 真诚的，真心的

14 enhance /ɪn'hɑːns/ v. 提高，增强
15 interact[★] /ˌɪntər'ækt/ vi. 互相作用，互相影响
16 regardless of[★] 不顾
17 mutual /'mjuːtʃuəl/ a. 彼此的

After Reading

A. Main Idea

Complete the following diagram with the sentences given below.

1. A good work climate can be created, which can result in the company's high productivity and the workers' high performance.
2. The leadership and workers share the same purpose and vision of the company.
3. The leadership and workers can adjust quickly to the changing market conditions.
4. People respect and support each other and like to work together.
5. People are willing to accept new ideas, to take risks and to adopt innovations.
6. People are trained to interact and work together.
7. The company cares for the needs of every worker.
8. People are encouraged to be creative.
9. Each worker is encouraged to do his best.

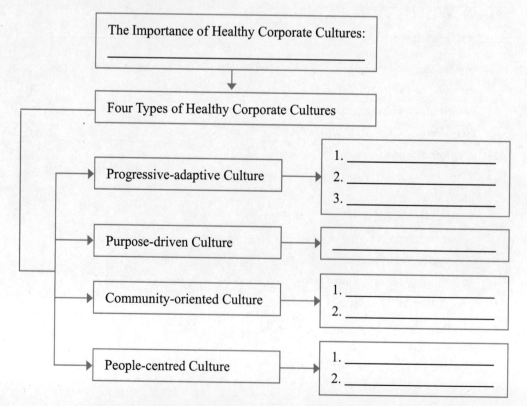

B. Detailed Understanding

I. Make correct statements according to the text by combining appropriate sentence parts in Column A with those in Column B.

Column A	Column B
1. Corporate culture can produce _____.	a. is encouraged to do as well as he/she possibly can
2. Progressive-adaptive culture means _____.	b. a larger area than team building
3. In a community-oriented culture, a company is _____.	c. a change in how the company is managed and how people are promoted
4. Purpose-driven culture means _____.	d. people are open to new ideas and willing to take risks and adopt innovations
5. People-centered culture refers to _____.	e. emphasizes a lot on cooperation
6. Community building covers _____ _____.	f. workers' creative spirit
7. In a people-centered culture, every worker _____.	g. a good work climate, which can result in the company's high productivity and the workers' high performance
8. The progressive-adaptive culture encourages _____.	h. like a genuine community, where workers are treated as important members
9. Good corporate culture can create _____.	i. the fact that a company really cares for every worker
10. Community-oriented culture _____ _____.	j. the leadership and workers share the same purpose and vision of the company

II. Explain the sentences by filling in the blanks.

1. **Text sentence:** Positive corporate cultures create a positive work climate, which can lead to productivity and job satisfaction.

   Interpretation: Good corporate cultures can _____.

2. **Text sentence:** They contribute to high performance without apparently linking reward to performance.

   Interpretation: They help _____.

3. **Text sentence:** There is openness to new ideas and a willingness to take risks and adopt innovations.

   Interpretation: People are open _____.

4. **Text sentence:** ... the only certainty it values is that the company is future-oriented and innovative.

   Interpretation: ... the only certain thing that is thought to be important is _____.

5. **Text sentence:** The management makes every effort to be on the cutting edge, and encourages continuous development of workers.

   Interpretation: The management tries hard _____.

6. **Text sentence:** The leadership communicates the purpose of the company effectively, so that there is a common purpose, a shared vision for all the workers.

   Interpretation: The leadership expresses _____.

7. **Text sentence:** Workers are highly motivated, because they are committed to the same set of core values.

   Interpretation: Workers are very willing _____.

8. **Text sentence:** Everyone is valued, regardless of their positions in the company.

   Interpretation: Everyone is important, _____.

9. **Text sentence:** The organization cares for the whole person in terms of recognizing workers, basic needs for learning and growth, for belonging and being connected.

   Interpretation: The organization is concerned about _____.

C. Detailed Study of the Text

----

1 **Positive corporate cultures create a positive work climate, which can lead to productivity and job satisfaction.** (Para. 2) 积极的企业文化能营造良好的工作氛围，并由此带来高生产率和员工对工作的满意度。

句中 positive 的意思为"好的；积极的；肯定的"。如：a positive answer 肯定的答复。

2 The management makes every effort to be on the cutting edge, and encourages continuous development of workers. (Para. 4) 公司的管理者努力使自己处于领先位置，鼓励员工们不断完善自己。

句中 cutting edge 意为"最前列"。

3 The leadership communicates the purpose of the company effectively, so that there is a common purpose, a shared vision for all the workers. (Para. 5) 领导层向员工们有效地传达公司的意图，使员工们拥有共同的目标和理想。

句中 so that 引导的是状语从句，意思为"以便"。

4 The leadership attempts to build a community, in which people respect and support each other, and enjoy working together. (Para. 6) 领导层努力去建立一种群体，在这个群体里，人们互相尊重、互相支持、乐于在一起工作。

句中 in which people... work together 为状语从句，which 指的是 community。

5 Everyone is valued, regardless of their positions in the company. (Para. 8) 每一位员工，不论其在公司中的职务大小，都受到尊重。

句中短语 regardless of 的意思为"不论，不顾"。如：I'm buying that book, regardless of the cost. 我打算买下那本书，不管花多少钱。

D. Further Work on the Text

Write down at least three more comprehension questions of your own. Work in pairs and ask each other these questions. If you can't answer any of these questions, ask your classmates or the teacher for help.

1. _____

2. _____

3. _____

E. Vocabulary Practice

I. Find the word that does NOT belong to each group.

1. A. productivity   B. innovation   C. interaction   D. satisfaction
2. A. innovative   B. adaptive   C. creative   D. restless
3. A. community building B. team building C. group performance D. individual performance
4. A. enhance   B. extensive   C. motivate   D. reflect

II. Complete the following sentences with appropriate words in their correct form.

1. **possess, possessive, possession, possessor**

   1) She is now the proud _____ of two satellite dishes.

   2) He is very protective and _____ of his new car.

   3) I don't think Tom _____ a suit.

   4) This house has been in the family's _____ since 1900s.

2. **character, characteristic(*a.*), characterize, characterless, characterization**

   1) Some snack-bars sell mass-produced, _____ food.

   2) I would _____ Captain Jack as a born leader of men.

   3) She writes exciting stories but her _____ is weak.

   4) George, with _____ generosity, invited us all back to his house.

   5) In only ten years the whole _____ of that school has changed.

3. **innovate, innovation, innovative, innovator**

   1) He is now working in a young _____ company.

   2) Do you know who is the _____ of this new technology?

   3) They must encourage _____ if the company is to remain competitive.

   4) That scientist _____ new ways of research.

4. **lead, leader, leading, leadership**

   1) There were lights to _____ him there.

   2) That organization lacks _____ and direction.

   3) His air of confidence makes him a natural _____.

   4) Cancer is the _____ killer among elders in that city.

5. **regard, regarding, regardless**

   1) _____ of danger, that boy climbed the tower.

   2) I must speak to you _____ this problem.

   3) He _____ his wife's behavior with amusement.

Word Study

define

v. 1. 解释，给……下定义：A dictionary defines words. 词典给单词下定义。

2. 规定：The powers of a judge are defined by law. 法官的权限是由法律规定的。

派 definition

consult

v.   1. 请教，向（专业人员）咨询：to consult a lawyer 请教律师

2. 查阅，查看：to consult a dictionary 查阅词典

派   consultation, consultant

adjust

v.   1. 调节，改变……以适应：He adjusted himself quickly to the heat of the country. 他很快适应了这个国家的高温。

2. 校正，调整：Check and adjust the brakes regularly. 要定期检查并调校刹车。

adjust to/adjust... to （使……）适应于：She adjusted well to Washington. 她很能适应华盛顿的生活。

develop

v.   1. 发展，扩展，使……发达：Developing the national economy is a priority of the new government. 新政府的当务之急是发展国民经济。

2. 使（胶卷等）显影：Did you get the films developed? 你有没有把那些胶卷冲洗出来？

3. 成长，发育：He has developed into a tall, strong man. 他已长成一个高个儿壮汉。

派   development, developer, developing, developed

reward

n.   报酬，赏金，奖赏：The fireman received a reward for saving the child's life. 因为救了小孩的命，那位消防员受到了奖励。

v.   1. 酬劳，奖赏：The organizers rewarded the winners with gifts of fruits and flowers. 组织者用水果和鲜花奖赏得胜者。

2. 报答：How can I reward your help? 我怎样才能报答你的帮助呢？

Writing Practice

I. Complete the following sentences with the word provided in brackets.

Example: <u>When would you say</u> we should meet? (say)

1. _____ be a good person to ask? (think)

2. _____ is a good time to arrive? (suggest)

3. _____ we should go for a good meal? (advise)

4. _____ I should do to lose weight? (recommend)

5. _____ is wrong with Bill? (suppose)

II. **Use a negative question as a response in each the following dialogs, making use of the words given in brackets.**

Example:

A: Can you show me where his office is?

B: Why? **Haven't** you been there before?

1   A: I'm afraid I won't be able to give you a ride to the airport.

     B: Why not? _____? (promise)

2   A: I've left my job at Harvard.

     B: Why? _____? (happy)

3   A: Could you please help me buy a telephone card?

     B: Why? _____? (know, yourself)

4   A: Maybe it would be better not to be too good to Jane.

     B: Why not? _____? (friend of yours)

5   A: I really hate his sitting in my class every week.

     B: Why? _____? (normal, teacher evaluation)

6   A: I'm sorry I don't know the answer.

     B: Why not? _____? (expert, subject)

7   A: I've been waiting for you for one hour.

     B: Why? _____? (message, late)

8   A: I've not been able to finish the work.

     B: Why Not? _____? (Jane, help, promise)

III. **Fill in the blanks with "Yes" or "No".**

Example: You're not an English teacher, are you? ➔ **_No_**, I teach Russian.

1   A: You're not a student, are you?

     B: _____, I am studying English and history.

2   A: Couldn't you come home early today?

     B: _____, I've got too much to do.

3   A: Don't you want to wait for the results?

     B: _____, I think I'll come back later.

4   A: Didn't you tell me that your uncle was a dentist?

     B: _____, he was a lawyer.

5   A: Wouldn't you like another coffee?

     B: _____, that would be lovely.

6 A: Didn't you see him at the station?

   B: _____, I didn't see him.

7 A: Can't you find somebody to go with you?

   B: _____, everyone is busy.

8 A: Didn't you drive your car to school?

   B: _____, there isn't any parking space for students.

IV. Rewrite the sentences according to the models.

Model A:

Original sentence: People are open to new ideas and willing to take risks and adopt innovations.

    New sentence: There is openness to new ideas and a willingness to take risks and adopt innovations. (Text B)

1. People are excited about the visiting football team.

    _____

2. People are full of creative energy, and are willing to accept new ideas.

    _____

Model B:

Original sentence: They help the employees to do their best, but apparently they don't link reward to performance.

    New sentence: They contribute to high performance without apparently linking reward to performance. (Text B)

3. He left the room, but he didn't tell me.

    _____

4. How dare you do such a thing? You didn't consult me.

    _____

Model C:

Original sentence: The owner of Company A became convinced that he was a big problem for his own company, and for the development of his company, he had to leave.

    New sentence: The owner of Company A became convinced he was such a problem that, for his business to grow, he had to leave it. (Text A)

5. He is an idiot. I won't ask him to help any more.

    _____

6. It's a very tiny kitchen. I don't have to do much to keep it clean.

    _____

Model D:

Original sentence: Everyone is valued, no matter what his position is in the company.

　　New sentence: Everyone is valued, regardless of his position in the company. (Text B)

7. Everyone is treated equally, no matter what his race, religion or sex is.

    _____

8. No matter what the danger was, he climbed the tower.

    _____

V. **Combine each set of the sentences into one, using the connective words or expressions provided.**

1. a. The owner of Company A became convinced.

    b. He was a problem for his company.

    c. In order to let his business grow, he had to leave his company.

    New sentence: _____ (such... that, for)

    　　(Text A)

2. a. They made mistakes.

    b. I made certain the mistakes were small.

    c. I encouraged them to learn.

    d. I encouraged them to move on.

    New sentence: _____ (but, and)

    　　(Text A)

3. a. The owner of Company B also acknowledged he was part of the problem.

    b. He decided to work harder.

    c. He created some problems.

    d. He decided he could overcome the problems.

    New sentence: _____ (also, but, that, by)

    　　(Text A)

4. a. Positive corporate cultures create a positive work climate.

    b. This climate can lead to productivity.

    c. It can also lead to job satisfaction.

    New sentence: _____ (which, and)

    　　(Text B)

5. a. The leadership communicates the purpose of the company effectively.

   b. As a result, there is a common purpose.

   c. There is a shared vision for all the workers.

   New sentence: _____ (so that)

      (Text B)

6. a. A community-oriented culture goes beyond team building.

   b. It seeks to create a genuine community.

   c. In this community, every worker is treated as a valuable member.

   New sentence: _____ (and, in which)

      (Text B)

7. a. The organization cares for the whole person.

   b. The organization recognizes workers' basic needs.

   c. The workers' basic needs are about learning and growth, belonging and being connected.

   New sentence: _____ (in terms of, for)

      (Text B)

8. a. The owner did not want to fire any minor employees.

   b. He had become a friend to them.

   c. He had become a "father-figure" to them.

   New sentence: _____ (because, and)

      (Text A)

VI. Translate the following sentences into English.

1. 这项工作移交给了秘书。（turn over）
2. 他可能会迟到，那样我们就该等他。（in which case）
3. 我们看电影看得入了迷，连时间都忘了。（catch up in）
4. 要适应那里的高温比他们预期的要难。（adjust to）
5. 宋朝为世界文明做出了三大发明的贡献。（the Song Dynasty, contribute... to）
6. 这些夜校课程读完可得学位。（lead to）
7. 他感到他不会得到进一步的提升。（hold back）
8. 这本书由10个章节组成。（make up）

VII. Translate the following paragraph into Chinese.

    Many articles and books have been written in recent years about culture in organizations, usually referred to as "Corporate Culture". Every organization has its own unique culture or value set. Most organizations don't consciously try to create a certain culture. The culture of the organization is typically created unconsciously, based on the values of the top management or the founders of an organization.

VIII. Practical English Writing

Directions: Study the management styles of the following two companies.

Company A	Company B
1. All decisions are made by the management.	1. Voices of the employees are heard before any decisions are made.
2. The owner cares much about the market.	2. The owner cares much about employees.
3. The owner believes that the management is made up of wise people.	3. The owner believes that all employees should be regarded as powerful and creative people.
4. When rules are made, they should be followed.	4. Rules are made by people, so they can be changed by people.
5. Efforts are made to hire more qualified employees.	5. Efforts are made to train employees to perform new tasks.
6. The management is interested in knowing how other companies are doing.	6. The management is interested in knowing whether everyone in the company is doing his/her best.

Now discuss with your classmates and decide which company you prefer. Then write a short passage giving the reasons for your choice.

I Prefer Working in Company A/B

_____

_____

_____

_____

_____

_____

5
Unit

 Text A

 Before Reading

Discuss the following questions in class.

1. What do you know about robots?

2. Do you think robots can take the place of man one day? If yes, in what way? If not, why not?

 Reading

The Robot Protests

1   **City-Cell[1] 1: Central, 9 August** Human police were called into action again today as domestic[2] robots continued to riot[3] in the streets of CC14. Sixteen humans have been admitted to hospital so far this month in the worst outbreak[4] of robot violence ever recorded.

2   *Central* was the scene[5] of the worst violence. More than 2,000 robots stormed Statue[6] Square and destroyed the bronze[7] figure of a human which had been placed there almost 150 years ago. They then took to the streets and pushed dozens of business humans from their laser-riders[8]. Two business humans were injured[9] and traffic was brought to a standstill[10] as far away as CC6: Peak[11] II as Hover-trams[12] were brought in to carry away the victims.

3   According to a spokesrobot[13], the protesters were angry about what they claimed[14] to be the unfair treatment[15] of robots. The robot, a model D-7 Kingtron that refused to give its name, told reporters at the scene that robots were tired of being treated like second-class citizens. The D-7 Kingtron said that they would no longer put up with[16] being thought of as slaves[17] for humans.

4   "We are assigned[18] to do all the dirty, dangerous and repetitive[19] jobs that humans don't like doing," it said. "We are highly advanced techno[20]-units, with the ability to do complex tasks requiring high intelligence. Why are we forced to clean up household rubbish, do the laundry[21] and look after pets?"

5   Other robots involved in the riot agreed. One model G-4 said it would gladly assist[22] its humans with household chores if they would only show some appreciation[23]. Another robot, a B-52, added that cooking and caring for human children was not so bad, but that it had to draw the line at cleaning windows. It said that it was humiliating[24] to be seen doing such trivial[25] chores by executive robots in nearby offices.

6 Police Chief[26] Inspector[27] Saturn Ho said that the situation was out of control and that they were considering contacting the Starship Luna for help in ending the riots. "These are clever robots," he said. "They know exactly what they're doing and they won't stop until they get what they want."

7 PCI Ho explained that the rioting robots had been programmed to see, hear and feel. He said there was no telling what these robots would do if their demands for equal status with humans were not met. He was worried that sooner or later[28], someone would be killed.

8 Meanwhile, a hospital spokesrobot, a member of the Professional Robots Union which is not involved in the protests, said that none of the injuries they had treated were serious. Most of the injured humans, it said, were suffering from[29] muscle strain[30] in their arms and backs from doing household chores themselves while their robots were out rioting.

New Words and Expressions

1 city-cell /'sɪtɪˌsel/ *n.* 市区

2 domestic* /dəʊ'mestɪk/ *a.* 家庭的，家用的

3 riot▲ /'raɪət/ *v.* （参加）闹事，骚乱

4 outbreak# /'aʊtbreɪk/ *n.* （战事、情感、火山等的）爆发

5 scene /siːn/ *n.* （事故、案件等的）地点，现场

6 statue* /'stætʃuː/ *n.* 塑像，雕像

7 bronze■ /brɒnz/ *a.* 青铜制的

8 laser-rider# /'leɪzəˌraɪdə(r)/ *n.* 激光骑行器

9 injure /'ɪndʒə(r)/ *v.* （尤指在事故中）使（人、动物）受伤、弄伤

10 standstill# /'stændstɪl/ *n.* 静止状态，停止

11 peak▲ /piːk/ *n.* 顶峰

12 hover-tram■ /'hɒvətræm/ *n.* 气垫汽车

13 spokesrobot# /'spəʊksˌrəʊbɒt/ *n.* 机器人发言人

14 claim /kleɪm/ *v.* 声称，断言

15 treatment▲ /'triːtmənt/ *n.* 对待，待遇，处理

16 put up with 忍受，忍耐

17 slave /sleɪv/ *n.* 奴隶，苦工

18 assign* /ə'saɪn/ *v.* （*to*）指派，选派

19 repetitive[#] /rɪ'petətɪv/ *a.* 重复的

20 techno- /'teknəʊ/ 表示 "技术"，"工艺" 的前缀

21 laundry /'lɔːndrɪ/ *n.* 待洗的衣服，洗好的衣服

22 assist[★] /ə'sɪst/ *v.* 帮助，协助

23 appreciation /ə,priːʃɪ'eɪʃən/ *n.* 感激，感谢

24 humiliate■ /hjuː'mɪlɪeɪt/ *vt.* 使蒙羞，羞辱，使丢脸

25 trivial■ /'trɪvɪəl/ *a.* 琐碎的，不重要的

26 chief /tʃiːf/ *a.* 主要的，首要的

27 inspector[#] /ɪn'spektə(r)/ *n.* 检查员，监察员，巡视员

28 sooner or later 迟早，早晚

29 suffer from 经受（疾病，不愉快之事等）

30 strain▲ /streɪn/ *n.* （因用力过度而）拉伤，扭伤

Proper Nouns

Kingtron /'kɪŋtrən/ 金斯乔恩 [机器人名]
Starship Luna 星际飞船卢娜 [星际飞船名]
Saturn /'sætən/ 萨图恩 [人名]

After Reading

A. Main Idea

Complete the following diagram with the sentences or expressions given below.

1. Human police were called into action again.

2. were brought in to carry away the victims

3. Human police were thinking about contacting the Starship Luna for help.

4. by themselves while doing household chores rather than by robots in the riots

5. being thought of as slaves for humans

6. being treated like second-class citizens

7. do all the dirty, dangerous and repetitive jobs

8. clean up household rubbish, do the laundry and look after pets

9. Humans do not show appreciation.

10. destroyed the bronze figure of a human

11. pushed dozens of business humans from their laser-riders

12. Traffic was brought to a standstill.

13. Two business humans were injured.

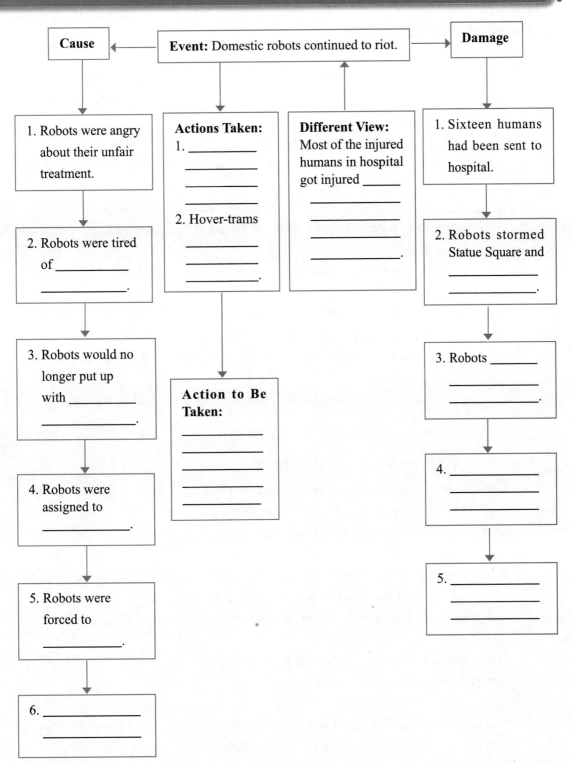

Cause ← **Event: Domestic robots continued to riot.** → **Damage**

1. Robots were angry about their unfair treatment.

2. Robots were tired of _____ _____.

3. Robots would no longer put up with _____ _____.

4. Robots were assigned to _____.

5. Robots were forced to _____.

6. _____ _____

Actions Taken:
1. _____
   _____
   _____
   _____
2. Hover-trams
   _____
   _____
   _____.

Action to Be Taken:
_____
_____
_____
_____
_____

Different View:
Most of the injured humans in hospital got injured _____
_____
_____
_____
_____.

1. Sixteen humans had been sent to hospital.

2. Robots stormed Statue Square and _____ _____.

3. Robots _____ _____ _____.

4. _____ _____ _____

5. _____ _____ _____

B. Detailed Understanding

- -

I. Tell if the following statements are true (T) or false (F) according to the text.

1. _____ People called the human police to report a riot of some robots who sent 16 humans to hospital.

2. _____ This riot was regarded as the worst violence so far caused by robots.

3. _____ Many robots launched an attack on Statue Square and destroyed a bronze figure of a human which had been there for almost 150 years.

4   _____ Two business humans were injured and traffic was stopped everywhere.

5. _____ Robots rioted because they could not stand being treated like second-class citizens.

6. _____ Robots would be willing to do whatever humans asked them to do if humans appreciated their work.

7. _____ Some robots thought what they had been doing was a waste of their talent.

8. _____ The police had got in touch with the Starship Luna for help.

9. _____ PCI Ho was worried that robots wouldn't stop rioting if their demands were not satisfied.

10. _____ According to a hospital spokesrobot, humans got injured by themselves rather than by robots.

II. Explain the sentences by filling in the blanks.

1. **Text sentence:** Human police were called into action again today as domestic robots continued to riot in the streets of CC14.

    Interpretation: Human police were ordered to _____.

2. **Text sentence:** Sixteen humans have been admitted to hospital so far this month in the worst outbreak of robot violence ever recorded.

    Interpretation: Up till now this month sixteen humans _____.

3. **Text sentence:** Central was the scene of the worst violence.

    Interpretation: The worst violence _____.

4. **Text sentence:** More than 2,000 robots stormed Statue Square and destroyed the bronze figure of a human which had been placed there almost 150 years ago.

    Interpretation: There were over 2,000 robots _____.

5. **Text sentence:** ... the protesters were angry about what they claimed to be the unfair treatment of robots.

   Interpretation: ... according to the claims of the robots, they protested _____.

6. **Text sentence:** The D-7 Kingtron said that they would no longer put up with being thought of as slaves for humans.

   Interpretation: According to the D-7 Kingtron, they were regarded as _____.

7. **Text sentence:** Other robots involved in the riot agreed.

   Interpretation: Other robots who _____.

8. **Text sentence:** He said there was no telling what these robots would do if their demands for equal status with humans were not met.

   Interpretation: He said it was impossible _____.

9. **Text sentence:** He was worried that sooner or later, someone would be killed.

   Interpretation: He was worried that _____.

C. Detailed Study of the Text

1 **Human police were called into action again today as domestic robots continued to riot in the streets of CC14.** (Para. 1) 今天人类警察再次受命采取行动, 因为从事家政的机器人继续在 CC14 地区的街上制造暴乱。

   call 意为 "要求, 命令", be called into action 意思是 "受命采取行动"。本句中 as 为连词, 引导原因状语从句。

2 **Sixteen humans have been admitted to hospital so far this month in the worst outbreak of robot violence ever recorded.** (Para. 2) 在本月爆发的这场有史以来最严重的机器人暴力事件中, 到现在为止已有 16 人被送进了医院。

   be admitted to hospital 意为 "被送进医院", 例如: He was admitted to a local hospital immediately after the accident. 事故发生后他立即被送进了医院。

3 **... but that it had to draw the line at cleaning windows.** (Para. 5) 但是它不得不拒绝擦窗户。

   句中 it 指代本句中的机器人。draw the line 意为 "划界线, (因不赞同而) 拒绝 (做某事)"。

4 **It said that it was humiliating to be seen doing such trivial chores by executive robots in nearby offices.** (Para. 5) 它说被附近办公室里从事管理工作的机器人看到它做着如此琐碎的家务是很丢人的事。

   句中第二个 it 为形式主语, 真正的主语为 to be seen doing such trivial chores by executive robots in nearby offices。

D. Talking About the Text

Work in pairs. Ask and answer the following questions first and then put your answers together to make an oral composition.

1. Why were human police called into action again today?

2. How did the riot compare with other outbreaks of robot violence? Did it cause any damage?

3. What did the robots do to Statue Square?

4. What happened to some humans during the riot? What happened to the traffic?

5. Why did the robots riot?

6. What kind of jobs were the robots asked to do?

7. What were the police going to do with the riot?

8. What was PCI Ho's concern?

9. What did a hospital spokesrobot think of the human injuries in this riot?

E. Vocabulary Practice

I. Fill in the blanks with the new words or expressions from Text A.

1. Within minutes, ambulances and police cars rushed to the _____ of the traffic accident.

2. The army was called in to put down the political _____ in the country.

3. The bird hopped around helplessly as if one of its wings had been _____.

4. One or two of them are so _____ that they really don't deserve attention.

5. They promised they would _____ the best man to the job.

6. He decided to quit his job because his boss _____ him in front of all his colleagues.

7. She _____ evil dreams a long time after that terrible accident.

8. _____ the same old routine, he decided to quit his job and start a new career of his own.

9. It was difficult to understand Mary's sudden _____ of anger.

10. A team of nurses _____ the doctor in performing the operation.

II. **Complete the following dialogs with appropriate words or expressions from Text A.**

1. A: The earthquake has so far _____ over 70,000 lives.

   B: It was really a horrible disaster.

2. A: I've been assigned the job of _____ the new students.

   B: Call me if you need more hands.

3. A: You always mess the room if I'm not here.

   B: I will remember to _____ it _____ next time.

4. A: You will be _____ to show your ID at the entrance.

   B: Thanks, I will remember to bring it with me.

5. A: I left the phone number of the local agent on your table in case you want to _____ him.

   B: That's very kind of you.

6. A: He gave up the experiment despite all his previous efforts.

   B: _____, he will regret it.

7. A: I didn't realize the project _____ so much work.

   B: That's why we started it a year earlier.

8. A: To show our _____ for all her hard work, I suggest holding a party this weekend.

   B: That's a good idea.

9. A: I will call to congratulate him on his _____ in the field of physics.

   B: So will I. How about sending him a card at the same time?

10. A: There was an accident during the rush hour this morning. I was told it was near your office.

    B: Yes. Actually a school bus was _____ and hit three stationary cars.

Text B

 Before Reading

Discuss the following questions in class.

1. What is science fiction?
2. Have you ever read any science fiction or seen any science fiction movie? If yes, what is it about ?

 Reading

Science Fiction[1] Shapes Our Ideas

1   In 1947, American author Robert A. Heinlein published a novel called *Rocket[2] Ship Galileo*, about a group of whiz[3] kids who build their own ship and fly into space.

2   This summer, 57 years after the book, a spacecraft[4] called SpaceShipOne was launched in the US, becoming the first manned space flight[5] by private citizens. The accomplishment capped[6] a remarkable story about a group of whizzes who decided one day to build their own ship and fly into space.

3   If the stories sound similar, it's because one inspired the other.

4   Science fiction became science fact. And now the stories of *Rocket Ship Galileo*, SpaceShipOne and the connection between the two occupy[7] the same display case as the newest exhibit in the Science Fiction Museum and Hall of Fame in Seattle, Washington, in the US.

5   The museum, which opened in June, is the US$20 million brainchild[8] of billionaire[9] sci-fi[10] fan Paul Allen. He happens to be one of the whizzes behind SpaceShipOne and the owner of a yellowed[11] paperback[12] copy of Heinlein's book. He read it when he was 11.

6   Science fiction is a big inspiration[13] for creativity and for thinking out of the box, Allen said in an e-mail exchange. It forces you to think about the world and about future possibilities, and it reinforces[14] the idea that creativity can be expressed in new ways through science and technology.

7   Advertised as the only one of its kind on the planet, the Science Fiction Museum traces[15] the evolution[16] of science fiction from its earliest pioneers in the 19th century through its transformation into a literature genre[17]. For many people and for many

years, science fiction was kid stuff.

8   But then something happened. Science fiction became not just respectable[18] but respected, according to Eric Rabkin, professor of English at the University of Michigan. Science fiction is now a mainstream genre.

9   "As recently as 10 years ago, people in the academic world thought of science fiction as beneath[19] consideration[20]", Rabkin said.

10   Today, major universities like Rabkin's offer courses in sci-fi literature. Science fiction writers now win MacArthur fellowships[21], the so-called "genius[22] grants" in the US. Five of the 10 best movies of all time are science fiction. *Star Wars: Episode[23] IV* and *E.T.* rank[24] second and fourth on the list.

11   A website called Technovelgy.com lists more than 675 inventions inspired or shaped by science fiction.

12   American physicist Leo Szilard, who first explained nuclear[25] chain reaction, got the idea from English novelist H. G. Wells' 1914 sci-fi book, *The World Set Free*, said Gregory Benford, a physicist at the University of California, Irvine, and author of a dozen sic-fi novels. Benford is an adviser to the museum. He said the term "atom bomb[26]" was invented by Wells in the same book.

13   The final seal[37] of the respectability has come from the group with the reputation as being the hardest to convince: scientists.

14   The museum's board of advisers includes not only writers and movie directors but accomplished physicists and aerospace[28] engineers. Two board members conduct research for the National Aeronautics and Space Administration (NASA). The museum director spent three decades as an aerospace engineer and led the NASA exploration programme.

New Words and Expressions

1 science fiction ▲ *n.* 科幻小说
2 rocket /'rɒkɪt/ *n.* 火箭
3 whiz /hwɪz/ *n.* 奇才
4 spacecraft * /'speɪskrɑːft/ *n.* 宇宙飞船
5 flight /flaɪt/ *n.* 飞行
6 cap /kæp/ *v.* 胜过，超过

7 occupy /'ɒkjʊpaɪ/ *vt.* 占，占据

8 brainchild# /'breɪntʃaɪld/ *n.* （口）脑力劳动的产物

9 billionaire /ˌbɪljə'neə(r)/ *n.* 亿万富翁

10 sci-fi /'saɪ'faɪ/ *n.* （非正式）科幻小说

11 yellowed /'jeləʊd/ *a.* 泛黄的，黄色的

12 paperback# /'peɪpəbæk/ *n.* 平装本，简装本

13 inspiration■ /ˌɪnspə'reɪʃən/ *n.* 给人以灵感的人（或事物）

14 reinforce★ /ˌriːɪn'fɔːs/ *v.* 进一步证实

15 trace★ /treɪs/ *vt.* 追溯，探索

16 evolution★ /ˌiːvə'luːʃən/ *n.* 发展，演变

17 genre■ /'ʒɒŋrə/ *n.* （文学、艺术等的）体裁，风格

18 respectable# /rɪ'spektəbl/ *a.* 可敬的，值得尊敬的

   respectability /rɪˌspektə'bɪlətɪ/ *n.* 可敬

19 beneath /bɪ'niːθ/ *prep.* 不值得，不足于

20 consideration★ /kənˌsɪdə'reɪʃən/ *n.* 考虑

21 fellowship■ /'feləʊʃɪp/ *n.* 研究员基金

22 genius★ /'dʒiːnjəs/ *n.* 天才

23 episode▲ /'epɪsəʊd/ *n.* （连续剧的）一集

24 rank★ /ræŋk/ *vi.* （在序列中）占特定等级

25 nuclear★ /'njuːklɪə(r)/ *a.* 核子的，核能的

26 atom bomb 原子弹

27 seal★ /siːl/ *n.* 印章

28 aerospace■ /'eərəʊspeɪs/ *a.* [只作定语] 航空与航天（空间）的

Proper Nouns

Robert A. Heinlein /'rɒbəteɪ'haɪnlaɪn/ 罗伯特·A. 海因莱因 [人名]

Galileo /ˌgælɪ'leɪəʊ/ 伽利略 [1564-1642，意大利数学家、物理学家和天文学家]

Eric Rabkin /'erɪkˌræbkɪn/ 埃里克·拉布金 [人名]

Michigan /'mɪʃɪgən/ 密歇根州 [美国州名]

MacArthur /mək'ɑːθə(r)/ 麦克阿瑟 [人名]

Leo Szilard /'liːəʊ'zɪlɑːd/ 雷奥·西拉德 [人名]

H. G. Wells /wels/ H. G. 韦尔斯 [人名]

Gregory Benford /'gregərɪ'benfəd/ 格列高利·本福德 [人名]

Irvine /'ɜːvɪn/ 欧文 [地名]

NASA （美国）国家航空和宇宙航行局

After Reading

A. Main Idea

Complete the following diagram with the sentences or expressions given below.

1. More than 675 inventions were inspired or shaped by science fiction.
2. offer courses in sci-fi literature
3. now win MacArthur fellowships
4. Five of the 10 best movies of all time are science fiction.
5. Science fiction was kid stuff.
6. helped inspire the launching of SpaceShipOne
7. got his idea of nuclear chain reaction from the sci-fi book *The World Set Free*
8. Science fiction was beneath consideration.

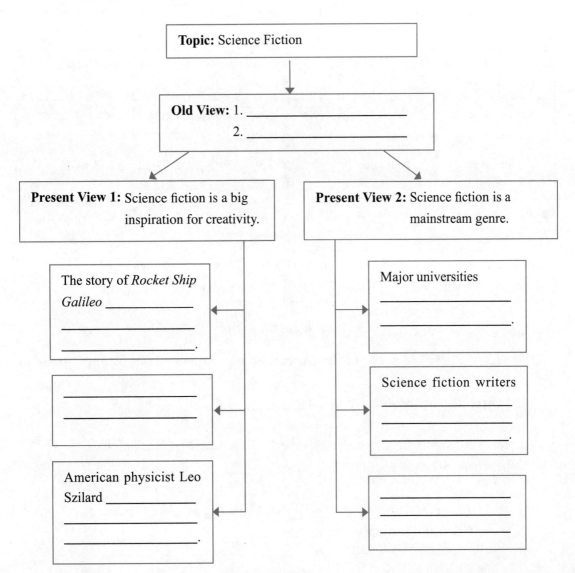

B. Detailed Understanding

I. Make correct statements according to the text by combining appropriate sentence parts in Column A with those in Column B.

Column A	Column B
1. *Rocket Ship Galileo* is a novel _____.	a. as "kid stuff" and something beneath consideration
2. Heinlein's novel *Rocket Ship Galileo* _____.	b. have been put on display in the Science Fiction Museum
3. SpaceShipOne was launched 57 years _____.	c. about a group of whiz kids who build their own ship and fly into space
4. Being the first manned spacecraft by private citizens, _____.	d. was shaped by a sci-fi book, *The World Set Free*
5. The stories of *Rocket Ship Galileo* and SpaceShipOne _____.	e. were both adapted from science fiction
6. The Science Fiction Museum is regarded as the only one of its kind _____.	f. because it traces the evolution of science fiction
7. Science fiction used to be treated _____.	g. SpaceShipOne embodied the ideas of *Rocket Ship Galileo*
8. The movies *Star War: Episode IV* and *E.T.* _____.	h. after the novel *Rocket Ship Galileo* was published
9. American physicist Leo Szilard's idea of nuclear chain reaction _____.	i. helped inspire the launching of SpaceShipOne

II. Explain the sentences by filling in the blanks.

1. **Text sentence:** The accomplishment capped a remarkable story about a group of whizzes...

   Interpretation: The accomplishment _____...

2. **Text sentence:** If the stories sound similar, it's because one inspired the other.

   Interpretation: The possible reason why _____.

3. **Text sentence:** He happens to be one of the whizzes behind SpaceShipOne...

   Interpretation: It so happened that _____...

4. **Text sentence:** Science fiction is a big inspiration for creativity and for thinking out of the box...

   Interpretation: Science fiction inspires not only creativity _____...

5. **Text sentence:** Science fiction became not just respectable but respected, according to Eric Rabkin...

   Interpretation: Eric Rabkin said that science fiction was something _____...

6. **Text sentence:** As recently as 10 years ago, people in the academic world thought of science fiction as beneath consideration...

   Interpretation: It has been only 10 years since _____...

7. **Text sentence:** A website called "Technovelgy.com" lists more than 675 inventions inspired or shaped by science fiction.

   Interpretation: According to a website called "Technovelgy.com", over 675 inventions _____.

8. **Text sentence:** The final seal of the respectability has come from the group with the reputation as being the hardest to convince: scientists.

   Interpretation: Even scientists who have been regarded as_____.

C. Detailed Study of the Text

1 **And now the stories of *Rocket Ship Galileo*, SpaceShipOne and the connection between the two occupy the same display case as the newest exhibit in the Science Fiction Museum and Hall of Fame in Seattle, Washington, in the US.** (Para. 4) 而现在,《伽利略火箭飞船》、太空飞船一号的故事以及二者之间的联系都被作为最新的展品陈列在位于美国华盛顿州西雅图市的科幻小说博物馆和名人纪念馆的同一展区。
华盛顿州位于美国西北部,西雅图是美国西部港口城市。

2 **Science fiction is a big inspiration for creativity and for thinking out of the box...** (Para. 6) 科幻小说不仅激发人们的创造力,而且使人们更具有创新思维能力……
think out of the box 也作 think outside the box,意为"思考新的、不同的东西,或不同的行为方式"。

3 **... and it reinforces the idea that creativity can be expressed in new ways through science and technology...** (Para. 6) ……它加深了这一观点,即创造性可以通过科技手段用新的方法表达出来……
it 指代前句中的 science fiction。that 所引导的从句为同位语从句,修饰 idea。

4   A website called "Technovelgy.com" lists more than 675 inventions inspired or shaped by science fiction. (Para. 11) 一个名为 Technovelgy.com 的网站列出了不下于 675 项发明，这些发明都受到了科幻小说的启发和影响。

called Technovelgy.com 为过去分词短语，在句中作定语，修饰 website。

inspired or shaped by science fiction 为过去分词短语，在句中作定语，修饰 inventions。

D. Further Work on the Text

Write down at least three more comprehension questions of your own. Work in pairs and ask each other these questions. If you can't answer any of these questions, ask your classmates or the teacher for help.

1. _____

2. _____

3. _____

E. Vocabulary Practice

I. Find the word that does NOT belong to each group.

1 A. fiction	B. novel	C. story	D. movie
2. A. writer	B. physicist	C. novelist	D. scientist
3. A. sound	B. wound	C. pound	D. round
4. A. museum	B. exhibit	C. whiz	D. art
3. A. inspiration	B. evolution	C. creation	D. comprehension

II. Complete the following sentences with appropriate words in their correct form.

1. accomplish, accomplished, accomplishment

1) Everyone longs for a sense of being _____.

2) He was regarded as a man who would never _____ anything.

3) Researchers have found that people find pride in _____.

2. evolve, evolution, evolutionary

1) In the course of _____, some animals became extinct on the earth.

2) It is now popularly believed that human beings _____ from apes.

3) The _____ process of living things aroused the students' interest.

3. **transform, transformable, transformation**

   1) My hometown has been _____ into a beautiful industrial city.

   2) The system has undergone a great _____ because of the new policy.

   3) A person's character is _____ in the process of his growth.

4. **respect (*n.*), respect (*v.*), respectable, respected, respectful**

   1) I could hardly believe that such a _____ lawyer was actually a thief.

   2) We must pay _____ to the needs of the old.

   3) He is a hardworking teacher, and the students _____ him a lot.

   4) All the kids are _____ to their parents.

   5) He was a highly _____ journalist before he retired.

5. **exhibit (*n.*), exhibit (*v.*), exhibition**

   1) The young painter has _____ his work in several galleries so far.

   2) There will be an _____ of black and white photographs at the Museum of Contemporary Art this week.

   3) Many _____ in the museum were donated by the local people.

Word Study

admit

v.　1. 承认，供认：He admitted having stolen the car. 他承认偷了那辆车。

　　2. 准许……进入，准许……加入：He was operated on as soon as he was admitted to the hospital. 他一进医院就动了手术。

派　　admission

assignt

v.　1. (*to*) 指派，选派：I was assigned to look after the new students. 我被派去照料新来的学生。

　　2. 分配；布置（作业）：Your teacher assigned your work to be done at home. 你们老师给你们布置了回家应做的作业。

　　3. 指定（时间、地点等）：Shall we assign Thursdays for our weekly meeting? 我们把每周例会定在星期四好吗？

派　　assignment

inspire

v.　1. 鼓舞：His success inspired us to work still harder towards our set goal. 他的成功激励着我们朝着既定的目标更加努力工作。

　　2. 激起：This behavior inspired dislike in me. 这种行为激起了我的厌恶。

　　3. 给……以灵感：The memory of his mother inspired his best poem. 对母亲的怀念给他灵感，使他写出自己最好的诗。

派　**inspiration**

claim

v.　1. 宣称：He claims to have succeeded but I don't believe him. 他自称成功了，但我不信。

　　2. 对……提出要求，索取：None of the passengers claimed the suitcase. 没有乘客认领这只箱子。

n.　1. 宣称：The government say they have reduced personal taxation, but I would dispute this claim. 政府声称已经减收了个人税，对于这种说法我持有异议。

　　2. 索赔：The company has accepted his claim for compensation. 公司已接受了他的索赔要求。

record

v.　1. 记录，登记：The event will be recorded in history. 那次事件将载入史册。

　　2. 录音：She has recorded several songs. 她已经录制了好几首歌。

n.　1. 记录，记载：Keep a record of how much you have spent. 把你所花的钱记个账。

　　2. 最高纪录，最佳成绩：She holds the world record in the 800-meter race. 她保持着 800 米跑世界纪录。

派　**recorder, recording**

Writing Practice

I. Complete the following sentences, using the correct forms of ellipsis or substitution.

Example: —Will you be late home tonight? —I'm afraid <u>so</u>.

1. —Do you think it will be some bad news? —I hope _____.
2. —Are there any letters for me? —No, I'm afraid _____.
3. —After living in a big city for so long, Helen longs for some quietness in the country.
   —I imagine _____.
4. —It's going to rain. —Yes, it seems _____.
5. —We'd better not borrow Joe's camera without asking him. —No, I suppose _____.
6. —Do you think she'll like this present? —I'm certain she _____.
7. —You'll be going to the New Year party, won't you? —I expect _____.
8. —Do you think she'd like to go on holiday with us? —I know she _____.
9. —Were they angry about our decision? —It certainly seemed _____.
10. —Has Jack won the race? —It appears _____.

II. Improve the following sentences where necessary.

1. I didn't think Dunaway knew Brian, but apparently he knows.

   _____

2. She felt capable of taking on the job and she was well qualified to do so.

   _____

3. —Shall we play tennis? —No, I don't want.

   _____

4. I have never met the President. I would welcome the opportunity of meeting the President.

   _____

5. —Can I have a look around the house? —Of course, go wherever you want to.

   _____

6. I thought the movie was really good, and Diane, too.

   _____

7. —Are you coming to the party tonight? —Well, I'm not sure I want.

   _____

8. —There is no need for you to help me wash up. —But I'd like to.

   _____

9. Amy's Chinese teacher told her that she must practice every day. She has practiced every day since then.

   _____

10. —You ought to ask Jim for help. —I know that, but I don't like.

    _____

11. I don't like going to the dentist. None of us in our family likes.

    _____

12. —Do you think the children would like to go to a boxing match? —I know they'd like but I don't think they're old enough.

    _____

III. Respond appropriately with sentences using "one" or "ones".

Example:

What happened after the secretary resigned? Was a new one appointed immediately?

1. I'd like to have a word with your daughter; I mean the younger _____, not the elder _____.

2. Do you know our neighbors, the _____ next door?

3. I love teaching children, especially small _____.

4. I've met that young man somewhere, the _____ you don't seem to like very much.

5. Two postmen deliver letters here, a black _____ and a white _____.

6. Which shoes are yours, the blue _____ or the yellow _____?

7. I'm spending the weekend going to a museum, the _____ near the town center, I mean.

8. The Presidential debates this year were not as exciting as the _____ last year.

IV. Rewrite the sentences according to the models.

Model A:

Original sentence: The reason why we find the stories similar is that one story inspired the other story.

New sentence: If the stories sound similar, it's because one inspired the other. (Text B)

1. The reason why the experiment failed is that they used a wrong method.

   _____

2. The reason why young people like pop music is that it suits the rhythm of modern life.

   _____

Model B:

Original sentence: The worst violence broke out in Central.

New sentence: Central was the scene of the worst violence. (Text A)

3. The traffic accident took place at the bridge's eastern entrance.

   _____

4. The riot broke out at the city center.

    _____

Model C:

Original sentence: The protesters were angry because they claimed that robots were unfairly treated.

New sentence: The protesters were angry about what they claimed to be the unfair treatment of robots. (Text A)

5. The teachers were proud because they claimed that the school had the first-rate equipment.

    _____

6. They were very sad because they claimed that they had made a careless mistake.

    _____

Model D:

Original sentence: We are highly advanced techno-units and we have the ability to do complex tasks that require high intelligence.

New sentence: We are highly advanced techno-units, with the ability to do complex tasks requiring high intelligence. (Text A)

7. They are very responsible and will be able to finish the work which requires great patience.

    _____

8. He is a good actor and he has the ability to act in plays which call for both experience and talent.

    _____

V. Combine each set of the sentences into one, using the connective words or expressions provided.

1. a. Human police were called into action again today.

    b. Domestic robots continued to riot in the streets of CC14.

    New sentence: _____ (as) (Text A)

2. a. More than 2,000 robots stormed Statue Square.

    b. They destroyed the bronze figure of a human.

    c. The bronze figure had been placed there almost 150 years ago.

    New sentence: _____ (and, which)

           (Text A)

3. a. Two business humans were injured.

    b. Traffic was brought to a standstill as far away as CC6.

    New sentence: _____ (and) (Text A)

4.   a. We are assigned to do all the dirty, dangerous and repetitive jobs.

    b. Humans don't like doing these jobs.

    New sentence: _____ (that) (Text A)

5.   a. One model G-4 said it would gladly assist its humans with household chores.

    b. They would only show some appreciation.

    New sentence: _____ (if) (Text A)

6.   a. Most of the injured humans, it said, were suffering from muscle strain in their arms and backs.

    b. The humans injured themselves from doing household chores themselves.

    c. Their robots were out rioting.

    New sentence: _____ (while)
        (Text A)

7.   a. The accomplishment capped a remarkable story about a group of whizzes.

    b. The group of whizzes decided one day to build their own ship.

    c. They also decided one day to fly into space.

    New sentence: _____ (who, and)
        (Text B)

8.   a. It forces you to think about the world.

    b. It forces you to think about future possibilities.

    c. It reinforces an idea.

    d. Creativity can be expressed in new ways through science and technology.

    New sentence: _____ (and, that)
        (Text B)

VI. Translate the following sentences into English.

1. 由于大雾，市中心的交通陷于瘫痪。（standstill）
2. 她的观点不只是可以接受，而是被普遍接受的。（not just... but）
3. 直到20年前，人们才认识到吸烟有害健康。（as recently as）
4. 在陌生的环境中，你必须忍受许多的不便。（put up with, inconvenience）
5. 当我们去拜访她的时候，她碰巧不在家。（happen to）
6. 这是有记载以来最强烈的地震。（record）
7. 我可以帮忙，但撒谎的事我不干。（draw the line at）
8. 体育比赛中，通常不会考虑年龄因素。（beneath consideration）

VII. Translate the following paragraph into Chinese.

    Newton was said to have told many people the following story. In 1666, while taking a rest in the garden at home, he saw an apple falling off the tree. This gave him the inspiration and evoked his thinking over the problem of whether the power that pulled the apple down to the ground and the one that makes the moon revolve around the earth were the same. And it finally led to the establishment of the Law of Gravity, one of the great discoveries that have the most far-reaching impact on the development of science.

VIII. Practical English Writing

Directions: Science has changed our lives in many different ways. For example, medical science has helped us live a much longer and healthier life than we used to, and physical science has enabled us to travel faster and in greater comfort. Now think about your life as students living on the college campus. Can you name some of the aspects of your life in which science has made a difference? Discuss with your classmates first and put down some notes in the following table. Some examples have been provided for your reference.

Influence of Science on Your Everyday Life	Influence of Science on Teaching and Learning
1. Improved toothbrush and toothpaste make our teeth bright and shining.	1. We can listen to English recordings with our MP3 players.
2. Refrigerators help us keep the milk fresh for several days.	2. We can do our homework on the computer and submit it via the Internet.
3.	3.
4.	4.
5.	5.
6.	6.
7.	7.
8.	8.
9.	9.

Now write a short passage summarizing what you have found about the influence of science on your daily life or on teaching and learning.

Influence of Science on _____

_____

_____

_____

_____

_____

_____

_____

Text A

 Before Reading

Discuss the following questions in class.

1. What do you know about terrorism?
2. Do you remember anything about the terrorist attack on New York City on September 11, 2001?

 Reading

Focusing on Anti-Terrorism[1]

1   OKLAHOMA CITY—Long before a truck bomb tore open a federal building here and jets[2] piloted[3] by terrorists brought down the World Trade Center, Dennis J. Reimer said he recognized that something had to be done about the growing threat[4] of terrorism.

2   As a 37-year Army veteran[5], he had seen international violence firsthand[6]. He spoke to Congress[7] in the late 1990s about the potential dangers of unconventional[8] warfare[9], and shortly after that he was among the first wave of military[10] officers who willingly[11] agreed to take an anthrax[12] vaccine[13], believing bioterrorism[14] was the next line of terrorist offence[15].

3   A few years ago, after Reimer retired as the Army's 33rd chief of staff, the native Oklahoman got his chance, as he says, "to put my money where my mouth is."

4   The retired general in April 2000 became the first director of the National Memorial[16] Institute for the Prevention of Terrorism (NMIPT), a federally[17] funded research and development organization founded as an outgrowth[18] of the April 19, 1995 bombing here.

5   "As an Okie, I was given an opportunity to come back and help my home state. I knew the great tragedy[19] in 1995, and I believed deeply the nation needed to be prepared," Reimer, 64, said during a recent interview in his downtown offices here. "Because of what happened here, we believe Oklahoma City is a place where you can bring serious people together and talk about the complex issues associated with battling[20] terrorism on U.S. soil."

6   The concept[21] for an anti-terrorism center to prepare government officials, rescue[22]

workers and the public for another domestic attack was formulated[23] shortly after Timothy J. McVeigh blew up[24] the Federal Building, killing 168 people. State leaders, survivors[25] and relatives of the victims wanted—as part of the memorial complex—a nonprofit[26] organization whose broad mission[27] is to "prevent terrorism on U.S. soil." To date, it has largely focused its research and efforts on preparing the "first responders" —rescue workers.

7   "Family members and survivors wanted an organization that looked to the future," Reimer said. The organization is funded through the Department of Homeland Security's Office for Domestic Preparedness.

8   Today, the NMIPT works with and funds various university research projects: a study for improving crisis[28] communication and protecting telephone networks in the event of an attack; a project evaluating[29] the long-term mental health effects of an attack; studies on the development of anthrax vaccines; and the development of appropriate protective[30] clothing for emergency responders.

9   So far, its impact[31] and reach has been broad, experts say.

New Words and Expressions

1   anti-terrorism# /ˌæntɪ'terərɪzəm/ n. 反击恐怖主义
2   jet /dʒet/ n. 喷气式飞机
3   pilot /'paɪlət/ v. 驾驶（飞机）
4   threat* /θret/ n. 威胁，恐吓
5   veteran* /'vetərən/ n. 老兵
6   firsthand# /'fɜːst'hænd/ ad. 直接地，第一手地
7   Congress* /'kɒŋgres/ n. 美国国会
8   unconventional# /ˌʌnkən'venʃənəl/ a. 非常规的
9   warfare■ /'wɔːfeə(r)/ n. 战争（状态）
10   military /'mɪlɪtərɪ/ a. 军事的，军用的
11   willingly /'wɪlɪŋlɪ/ ad. 自愿地
12   anthrax# /'ænθræks/ n. 炭疽（牛羊的传染病，常可致命，也可传染给人）
13   vaccine■ /'væksiːn/ n. 疫苗，菌苗
14   bioterrorism# /ˌbaɪəʊ'terərɪzəm/ n. 生化恐怖
15   offence* /ə'fens/ n. 犯法行为，罪行
16   memorial* /mɪ'mɔːrɪəl/ a. 纪念性的，用以纪念的

17 federally# /'fedərəlɪ/ *ad.* 由联邦政府，通过联邦政府

18 outgrowth# /'aʊtɡrəʊθ/ *n.* 产物，后果

19 tragedy ▲ /'trædʒɪdɪ/ *n.* 惨事，灾难

20 battle /'bætl/ *v.* （与……）作战，斗争

21 concept /'kɒnsept/ *n.* 概念

22 rescue ★ /'reskjuː/ *n.* 营救，援救

23 formulate ■ /'fɔːmjʊleɪt/ *vt.* 构想出（计划、方法等）

24 blow up 爆炸

25 survivor# /sə'vaɪvə(r)/ *n.* 幸存者，生还者

26 nonprofit# /nɒn'prɒfɪt/ *a.* 非营利的

27 mission ★ /'mɪʃən/ *n.* 使命，任务

28 crisis /'kraɪsɪs/ *n.* 危机，危急关头

29 evaluate ★ /ɪ'væljʊeɪt/ *v.* 评价，估价

30 protective ▲ /prəʊ'tektɪv/ *a.* 保护的，防护的

31 impact ▲ /'ɪmpækt/ *n.* 影响，作用

Proper Nouns

Oklahoma City 俄克拉何马城［美国俄克拉何马州首府］

Dennis J. Reimer /'denɪsdʒeɪ'riːmə(r)/ 丹尼斯・J. 赖默 ［人名］

Okie /'əʊkɪ/ *n.* （美俚） 俄克拉何马州人

Timothy J. McVeigh /'tɪməθɪdʒeɪ,mək'veɪ/ 蒂莫西・J. 麦克维 ［人名］

 ## After Reading

A. Main Idea

Complete the following diagram with the expressions given below.

1. to prevent terrorism on U.S. soil

2. he spoke to Congress about the potential dangers of unconventional warfare

3. a study for improving crisis communication and protecting telephone networks in the event of an attack

4. became the first director of the National Memorial Institute for the Prevention of Terrorism

5. the development of appropriate protective clothing for emergency responders

6. he was among the first wave of military officers who willingly agreed to take an anthrax vaccine

7. a project evaluating the long-term mental health effects of an attack

8. federally funded research and development organization founded as an outgrowth of the April 19, 1995 bombing
9. studies on the development of anthrax vaccines

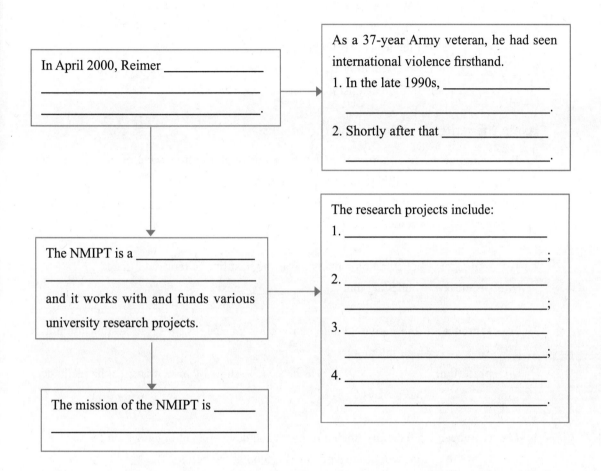

In April 2000, Reimer _____ _____ _____ .

As a 37-year Army veteran, he had seen international violence firsthand.
1. In the late 1990s, _____ _____ .
2. Shortly after that _____ _____ .

The NMIPT is a _____ _____ and it works with and funds various university research projects.

The research projects include:
1. _____ _____ ;
2. _____ _____ ;
3. _____ _____ ;
4. _____ _____ .

The mission of the NMIPT is _____ _____

B. Detailed Understanding

I. Tell if the following statements are true (T) or false (F) according to the text.

1. _____ Reimer served in the U.S. Army for 37 years.
2. _____ As soon as the World Trade Center was brought down by terrorists, Reimer realized that they had to do something to fight against terrorists.
3. _____ The aim of the NMIPT is to prevent terrorism in Oklahoma City.
4. _____ Timothy J. McVeigh was a criminal who was responsible for the bombing of the Federal Building in Oklahoma City.
5. _____ Rescue workers are supposed to respond as quickly as possible after a terrorist attack.

6. _____ The organization NMIPT was established shortly after the Federal Building in Oklahoma City was destroyed.

7. _____ Family members and survivors of the Federal Building bombing set up the NMIPT.

8. _____ Reimer used to be a high-ranking officer in the U.S. Army.

9. _____ Rescue workers need to wear special clothes in an emergency in order to protect themselves from being hurt.

10. _____ Because Oklahoma City experienced the tragedy, many people here were ready to fight against terrorism.

II. Explain the sentences by filling in the blanks.

1. **Text sentence:** ... and shortly after that he was among the first wave of military officers who willingly agreed to take an anthrax vaccine.

   Interpretation: ... and soon after that he became _____ to take an anthrax vaccine.

2. **Text sentence:** ... believing bioterrorism was the next line of terrorist offence.

   Interpretation: ... believing that _____ next.

3. **Text sentence:** ... the native Oklahoman got his chance, as he says, "to put my money where my mouth is."

   Interpretation: ... _____, as he says, to show by his actions that he really does what he says.

4. **Text sentence:** ... a federally funded research and development organization founded as an outgrowth of the April 19, 1995 bombing here.

   Interpretation: ... a research and development organization that was funded by the federal government and _____ Oklahoma City on April 19, 1995.

5. **Text sentence:** The concept for an anti-terrorism center to prepare government officials, rescue workers and the public for another domestic attack was formulated shortly after Timothy J. McVeigh blew up the Federal Building.

   Interpretation: Shortly after Timothy J. McVeigh destroyed the Federal Building, a plan was made to set up an anti-terrorism center to _____.

6. **Text sentence:** State leaders, survivors and relatives of the victims wanted—as part of the memorial complex—a nonprofit organization whose broad mission is to prevent terrorism on U.S. soil.

   Interpretation: State leaders, survivors and relatives of the victims believed that part of

the work in setting up the memorial complex should be the formation of
an organization _____.

7. **Text sentence:** Family members and survivors wanted an organization that looked to the
future.

   Interpretation: Family members and survivors _____.

8. **Text sentence:** Today, the NMIPT works with and funds various university research
projects.

   Interpretation: At present the NMIPT _____.

9. **Text sentence:** So far, its impact and reach has been broad.

   Interpretation: Up till now, _____.

C. Detailed Study of the Text

1　**Long before a truck bomb tore open a federal building here and jets piloted by terrorists
brought down the World Trade Center, Dennis J. Reimer said he recognized that
something had to be done about the growing threat of terrorism. (Para. 1)** 远在汽车炸
弹炸了这里的一幢联邦政府大楼以及恐怖分子驾驶飞机撞毁世界贸易中心之前，丹尼
斯·J. 赖默就说过，他意识到应采取措施以应付日趋严重的恐怖主义威胁。

World Trade Center: 世界贸易中心，建成于 1972 年。由南、北两座 110 层高的大厦组
成，曾是世界第二高楼。1993 年 2 月，恐怖分子对其实施爆炸式袭击，造成 6 人死亡。
2001 年 9 月 11 日上午 9 点左右，恐怖分子劫持了两架美国航空公司的客机，先后撞击
了北、南两楼，使世界贸易中心成为一堆废墟，制造了骇人听闻的"9·11"事件。

2　**A few years ago, after Reimer retired as the Armys 33rd chief of staff, the native
Oklahoman got his chance, as he says, "to put my money where my mouth is." (Para. 3)**
几年前，作为美国陆军第 33 任参谋长的赖默退役以后，这个俄克拉何马人正像其所说
的那样，"得到了兑现自己诺言的机会"。

chief of staff: 陆军参谋长，为美军参谋长联席会议的一员，也是总统陆军问题的主要军
事顾问。丹尼斯·赖默在 1995—1999 年期间任此职，因此，他有责任就国家安全问题
向国会发表意见。

to put one's money where one's mouth is 意为"说话算数，兑现承诺"。

3　**"... I knew the great tragedy in 1995, and I believed deeply the nation needed to be
prepared," Reimer, 64, said during a recent interview in his downtown offices here. (Para. 5)**
……64 岁的赖默在他位于市中心的办公室接受的一次采访中说："我知道 1995 年的

那场灾难，我深信这个国家有必要做好准备。"

此句中的 here 指的是俄克拉何马城。

4 **The concept for an anti-terrorism center to prepare government officials, rescue workers and the public for another domestic attack was formulated shortly after Timothy J. McVeigh blew up the Federal Building, killing 168 people.** (Para. 6) 在蒂莫西·J. 麦克维炸毁联邦大楼，导致 168 人丧生后不久，建立一个反恐中心的设想就被提了出来。这个组织的目的是培训政府官员，训练营救人员，使公众对新的国内袭击事件做好准备。

Timothy J. McVeigh: 蒂莫西·J. 麦克维，1995 年 4 月 19 日俄克拉何马联邦大楼爆炸案的两名主犯之一。2001 年 6 月 11 日被判处死刑。他是美国 38 年以来第一个被判死刑的犯人。

5 **The organization is funded through the Department of Homeland Security's Office for Domestic Preparedness.** (Para. 7) 这个机构是通过国土安全部的国内预警办公室来拨款的。

Department of Homeland Security (DHS): (美国) 国土安全部，它是 "9·11" 事件的产物，是 2002 年 11 月 25 日由美国总统布什签署成立的联邦行政机构。它由海岸警卫队、移民和归化局及海关总署组成。国土安全部的重要职责是保卫国土安全及相关事物，使美国能够更加有效地对付恐怖袭击威胁。

Office for Domestic Preparedness (ODP): 国内预警办公室，是国土安全部的一个主要部门。它的主要职责是为预防和应对恐怖事件而提供培训、拨款以及技术援助。

D. Talking About the Text

Work in pairs. Ask and answer the following questions first and then put your answers together to make an oral composition.

1. What did Dennis J. Reimer realize before the 9·11 attack and the Oklahoma City truck bombing?

2. What did Reimer speak about to Congress in the late 1990s?

3. Why did Reimer agree to take an anthrax vaccine?

4. What did Reimer become in April 2000?

5. Why did Reimer believe that Oklahoma City people could be brought together to talk about problems related with terrorism?

6. When did people plan to establish an anti-terrorism center?

7. What kind of organization was the center?

8. Through what organization is the NMIPT funded?

9. What do the university research projects funded by the NMIPT mainly deal with?

E. Vocabulary Practice

I. **Fill in the blanks with the new words or expressions from Text A.**

1. The ceremony was attended by many of the surviving _____ of World War II.
2. Despite his _____ methods, he has inspired students more than anyone else.
3. He was jailed for trying to _____ a plane.
4. Her first book is an _____ of an art project she began in 1988.
5. I would _____ help you if I weren't going away tomorrow.
6. She is without question the craziest person I've met _____.
7. Little by little he _____ his plan to escape.
8. If it's a choice between higher pay and job _____, I'd prefer to keep my job.
9. Most of the older reporters have experienced war _____.
10. These hospitals are _____ organizations and get tax relief.

II. **Complete the following dialogs with appropriate words or expressions from Text A.**

1. A: What's your _____?
   B: It is to isolate the enemies by destroying all the bridges across the river.
2. A: The doctor said he had passed the _____, and the fever was going down.
   B: Then we can have a good sleep tonight.
3. A: Are you all going to the _____ service held for Professor Conner?
   B: Sure, he was highly respected by us students. We will remember him forever.
4. A: The crime rate in that poor country has kept high for years.
   B: I believe crime is often an _____ of poverty.
5. A: What are these?
   B: There are _____ clothes which can reduce the absorption of chemicals through the skin.
6. A: Hitler's invasion of Poland led to the _____ of the Second World War.
   B: Hope such things would not be repeated in the future.
7. A: Do you think this product will sell well?
   B: The market situation is difficult to _____.
8. A: The disease has widely spread in that region, and children can be easy victims.
   B: Don't worry. Seven million doses of _____ have been delivered there.
9. A: He has to do his _____ service before going to university.
   B: That's the tradition of his country where they hope everyone will be trained like a soldier.
10. A: While the killer goes free he is a _____ to everyone in the town.
    B: So the police should take measures to catch him as soon as possible.

Text B

 Before Reading

Discuss the following questions in class.

1. Do you think guns are closely related with crimes? Why or why not?
2. Do you agree with the saying that weapons are threats to world peace?

 Reading

Too Many Guns?

1　　In August 1998, Chicago police officer Michael Ceriale was shot with a Smith & Wesson revolver[1] while on duty at a public housing project. He died a week later. His family blamed Smith & Wesson.

2　　If that strikes you as unreasonable, you're probably not a judge on the Illinois Appellate[2] Court. On Dec. 31, a three-judge panel[3] of the court unanimously[4] ruled that Ceriale's family and the families of four other victims could sue[5] the companies that produced the guns used to kill their relatives.

3　　Briefly put, the panel argue that gun makers make too many guns. Since criminals[6] use some of them, they say, this overproduction creates a "threat to the public". In essence[7], they say, firearm[8] makers "oversupplied" dealers[9] in Chicago's suburbs, knowing that some of the weapons would end up[10] in the city, where handguns are banned[11]. "Manufacturers[12] are endangering the public health and safety by knowingly[13] and intentionally supplying the criminal market."

4　　That certainly sounds bad, but it's hard to see how gun makers can be expected to stop their products from falling into the hands of[14] criminals. Let's assume they can figure out[15] what the reasonable demand ought to be in a given area and restrict[16] their shipments[17] to dealers accordingly[18]. Such arrangement plainly would not be enough to prevent criminals from shooting people. The gun fired by Ceriale's killer was originally shipped to a Houston dealer in 1980. No conceivable[19] precaution[20] by Smith & Wesson could have stopped it from being used to murder him nearly two decades later.

5　　"The manufacturers set in motion[21] a chain of events with the result being the death of our clients," said Jonathan Baum, an attorney[22] for the Ceriale family. While it's true that gun makers must know that some of their products will be used to commit murder, the only way to avoid that result is to go out of business.

6   But gun makers are not the only ones who should be worried. The same logic[23] could also be used against manufacturers of any product used by criminals. Doesn't Chrysler know that its cars will be used in bank robberies? Can't Rawlings anticipate[24] that gangsters[25] will use its baseball[26] bats to deliver vicious[27] beatings?

7   All these companies, and many others, could be said to "oversupply" their markets, making more of their products than are needed for legal[28] uses. If suing them seems more ridiculous[29] than suing Smith & Wesson, perhaps it's because guns are widely seen as less legal than cars and baseball bats.

8   Yet guns, like these other products, are used for lawful purposes far more often than they are used in crimes. And since one of these purposes is self-defense—for which they are used some two million times a year, according to national surveys—restricting the supply of firearms could cost lives rather than save them.

New Words and Expressions

1 revolver /rɪ'vɒlvə(r)/ *n.* 左轮手枪

2 appellate /ə'pelət/ *a.* 上诉的

3 panel /'pænəl/ *n.* 陪审团，全体陪审员

4 unanimously /juː'nænɪməslɪ/ *ad.* 一致地，无异议地

5 sue /sjuː/ *v.* 控告，上诉

6 criminal /'krɪmɪnəl/ *n.* 罪犯，犯人

7 in essence 本质上，实质上

8 firearm# /'faɪɑːm/ *n.* [常作 ~s] （便携式）枪支（尤指手枪和步枪）

9 dealer★ /'diːlə(r)/ *n.* 商人

10 end up 结束，告终

11 ban▲ /bæn/ *v.* 取缔，查禁

12 manufacturer▲ /ˌmænjʊ'fæktʃərə(r)/ *n.* 制造商

13 knowingly# /'nəʊɪŋlɪ/ *ad.* 故意地，蓄意地

14 fall into the hands of sb. 落到某人手里，成为某人所有

15 figure out 想出，弄清……的原因

16 restrict /rɪ'strɪkt/ *vt.* 限制，约束，限定

17 shipment★ /'ʃɪpmənt/ *n.* 装载（或运输）的货物

18 accordingly▲ /ə'kɔːdɪŋlɪ/ *ad.* 照着，相应地

19 conceivable▲ /kən'siːvəbl/ *a.* 可想到的，可想象的

20 precaution ★ /prɪˈkɔːʃən/ *n.* 防备，警惕

21 set sth. in motion 使某事物开始，导致某事发生

22 attorney ▲ /əˈtɜːnɪ/ *n.* 律师

23 logic ★ /ˈlɒdʒɪk/ *n.* 逻辑（学），逻辑性

24 anticipate ★ /ænˈtɪsɪpeɪt/ *vt.* 预料

25 gangster ■ /ˈɡæŋstə(r)/ *n.* 匪徒，歹徒

26 baseball ▲ /ˈbeɪsbɔːl/ *n.* 棒球

27 vicious /ˈvɪʃəs/ *a.* 恶意的，恶毒的

28 legal /ˈliːɡəl/ *a.* 合法的，法定的

29 ridiculous /rɪˈdɪkjʊləs/ *a.* 可笑的，荒谬的

Proper Nouns

Michael Ceriale /ˈmaɪkəlˈserɪəl/ 迈克尔·塞瑞尔［人名］

Smith & Wesson 史密斯·韦森枪械公司

Illinois /ˌɪlɪˈnɔɪ(z)/ （美国）伊利诺伊州［美国州名］

Houston /ˈhjuːstən/ 休斯敦［美国得克萨斯州东南部港市］

Jonathan Baum /ˈdʒɒnəθən bɔːm/ 乔纳森·鲍姆［人名］

Chrysler /ˈkraɪslə(r)/ （美国）克莱斯勒汽车制造公司

Rawlings /ˈrɔːlɪŋz/ （美国）罗林棒球公司

 After Reading

A. Main Idea

Complete the following diagram with the sentences or expressions given below.

1. Ceriale's family and the families of four other victims could sue the companies that produced the guns used to kill their relatives.

2. Gun makers make too many guns.

3. Chicago police officer Micheal Ceriale was shot with a Smith & Wesson revolver while on duty and died a week later.

4. Guns are used for lawful purposes far more often than they are used in crimes.

5. The manufacturers of weapons set in motion a chain of events with the result being the death of innocent people.

6. Manufacturers are endangering the public health and safety by knowingly and intentionally supplying the criminal market.

7. It's hard to see how gun makers can be expected to stop their products from falling into

the hands of criminals.

8. If gun makers are to be sued, manufacturers of any other products used by criminals should be sued too, because they also make more of their products than are needed.

9. to sue the companies that produced guns

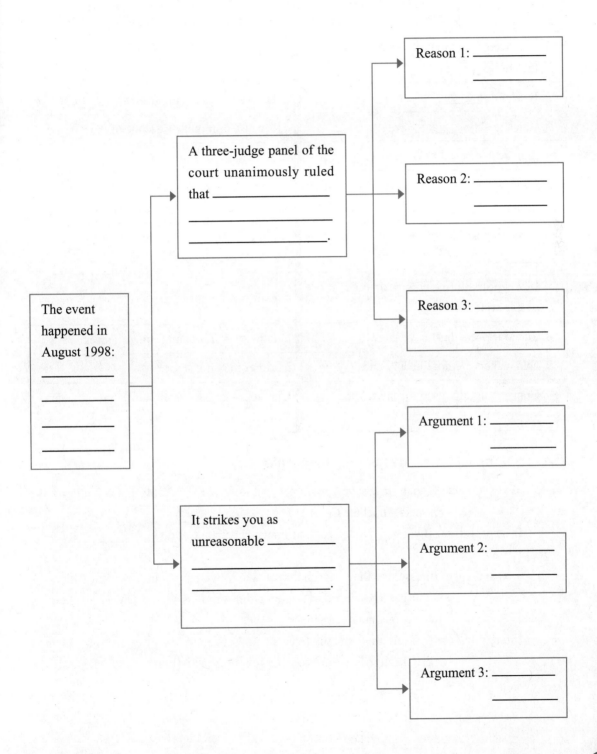

B. Detailed Understanding

I. Make correct statements according to the text by combining appropriate sentence parts in Column A with those in Column B.

Column A	Column B
1. The overproduction of weapons endangers the public _____.	a. even if it was bought several decades ago
2. It is hard to _____.	b. because guns are often used for self-defense
3. Chicago City _____.	c. because criminals use some of the weapons
4. A gun can be used to kill people _____.	d. to be used to commit murder
5. Even when all gun makers go out of business _____.	e. who should be worried about being sued
6. It is also possible for cars and baseball bats _____.	f. some of their products are supplying the criminal markets
7. If weapon supply was restricted, more people would die _____.	g. bans handguns
8. Manufacturers must have known _____.	h. rather than for committing offences
9. Gun makers are not the only ones _____.	i. crimes may not stop
10. Guns are usually used for lawful purposes, _____.	j. prevent weapons from falling into the hands of criminals

II. Explain the sentences by filling in the blanks.

1. **Text sentence:** If that strikes you as unreasonable, you're probably not a judge on the Illinois Appellate Court.

    Interpretation: If you are a judge on the Illinois Appellate Court, _____.

2. **Text sentence:** In essence, they say, firearm makers oversupplied dealers in Chicago's suburbs, knowing that some of the weapons would end up in the city, where handguns are banned.

    Interpretation: In fact, they mean, firearm makers provide the gun businessmen in Chicago's suburbs with too many of their products, and they _____.

3. **Text sentence:** Manufacturers are endangering the public health and safety by knowingly and intentionally supplying the criminal market.

    Interpretation: It is the intention of gun producers to _____ to the public health and safety.

4. **Text sentence:** Let's assume they can figure out what the reasonable demand ought to be in a given area and restrict their shipments to dealers accordingly.

    Interpretation: Suppose they can find out the appropriate _____.

5. **Text sentence:** Such arrangement plainly would not be enough to prevent criminals from shooting people.

    Interpretation: Obviously, such action _____.

6. **Text sentence:** No conceivable precaution by Smith & Wesson could have stopped it from being used to murder him nearly two decades later.

    Interpretation: Because the gun was sold 20 years ago, _____.

7. **Text sentence:** The manufacturers set in motion a chain of events with the result being the death of our clients...

    Interpretation: The gun production caused _____.

8. **Text sentence:** If suing them seems more ridiculous than suing Smith & Wesson, perhaps it's because guns are widely seen as less legal than cars and baseball bats.

    Interpretation: If it is more ridiculous _____, maybe it is because it seems more legal to use cars and baseball bats than to use guns.

9. **Text sentence:** And since one of these purposes is self-defense... restricting the supply of firearms could cost lives rather than save them.

    Interpretation: Since one of the things we do with guns is to protect ourselves... _____.

C. Detailed Study of the Text

1   **In August 1998, Chicago police officer Michael Ceriale was shot with a Smith & Wesson revolver while on duty at a public housing project. (Para. 1)** 1998 年 8 月，芝加哥警官迈克尔·塞瑞尔在一公共居民区值勤时遭到枪击，凶器是一把史密斯·韦森左轮手枪。
Smith & Wesson: 史密斯·韦森枪械公司，是美国最大的手枪制造公司。其总部位于马萨诸塞州，从 1852 年创建到现在已有 150 多年的生产历史。 housing project 是指公家出资营造、供低收入家庭居住的居民区或居民村。

2 **If that strikes you as unreasonable, you're probably not a judge on the Illinois Appellate Court.** (Para. 2) 如果你觉得这个结论不合理的话，你决不可能是伊利诺伊州上诉法院的法官。

Appellate Court: 上诉法院，它是在联邦法院 (federal courts) 中仅低于最高法院 (Supreme Court) 的一级审判机构。

3 **No conceivable precaution by Smith & Wesson could have stopped it from being used to murder him nearly two decades later.** (Para. 4) 史密斯·韦森枪械公司所能想出的任何预防措施都不能阻止 20 年后这把手枪被用来杀人。

it 指用来枪杀警官塞瑞尔的那把左轮手枪。him 指塞瑞尔。

4 **Doesn't Chrysler know that its cars will be used in bank robberies?** (Para. 6) 难道克莱斯勒公司不知道它的汽车可以用来抢劫银行吗？

Chrysler 克莱斯勒汽车制造公司，是美国第三大汽车工业公司，由沃尔特·克莱斯勒于 1923 年创建。它是一家跨国公司，总部设在底特律。

5 **Can't Rawlings anticipate that gangsters will use its baseball bats to deliver vicious beatings?** (Para. 6) 难道罗林棒球公司料想不到歹徒们会用它的球棒进行残暴的袭击吗？

Rawlings: 罗林棒球公司，是 1887 年由乔治·罗林兹和阿尔弗雷得·罗林兹创建的体育用品零售商店。现在，其商品已成为全美棒球联赛、棒球联合总会的指定用品。同时它也是全国大学生体育协会的橄榄球冠军赛、篮球冠军赛的指定供货商。

D. Further Work on the Text

Write down at least three more comprehension questions of your own. Work in pairs and ask each other these questions. If you can't answer any of these questions, ask your classmates or the teacher for help.

1. _____
2. _____
3. _____

E. Vocabulary Practice

I. **Find the word that does NOT belong to each group.**

1. A. legal   B. reasonable   C. ridiculous   D. logic
2. A. dealer   B. attorney   C. sue   D. panel
3. A. gangster   B. baseball   C. crime   D. criminal
4. A. revolver   B. weapon   C. manufacturer   D. firearm
5. A. unanimously   B. knowingly   C. conceivable   D. accordingly

II. **Complete the following sentences with appropriate words in their correct form.**

1. **housing, household, house**

   1) The _____ in this part of the city is bad.

   2) She lives in a small _____ .

   3) Children should learn to do some _____ chores.

2. **threatening, threaten, threat**

   1) The _____ of flood has been relieved.

   2) The company _____ with going out of business now.

   3) She received a _____ letter yesterday.

3. **ridicule, ridiculous, ridiculously**

   1) Silly mistakes and strange clothes often make someone the _____ of others.

   2) She looked _____ young to be a mother.

   3) It's _____ to expect a two-year-old to be able to read.

4. **crime, criminal (*a., n.*)**

   1) To spend your time in vain is a _____ waste of your life.

   2) Any _____ must be tried.

   3) It's a _____ to waste our country's natural resources.

5. **essence, essential, essentially**

   1) It's been believed for centuries that great writers and scientists are _____ different from ordinary people.

   2) The _____ of his argument was that education should continue throughout one's life.

   3) Government support will be _____ if the project is to succeed.

Word Study

blow

v.   1. 吹，吹动：The wind was blowing heavily during the voyage. 航行中风刮得很大。/He blew the dust off the table. 他吹掉桌上的灰尘。

2. 吹气，充气：Take a deep breath and blow. 深深地吸口气，再吹气。/The little girl blew the fire to flame. 小女孩把火吹旺了。

3. 吹响：Whistles blew, and the train started. 汽笛鸣响，火车开动了。

4. 爆炸，炸坏（某物）：The safe had been blown by the thieves. 保险箱被窃贼炸开了。

n.   一击，打击，捶打：He gave his opponent a violent blow on the back. 他在对手的背上狠击了一下。/His father's death was a fatal blow to him. 父亲的去世对他是个致命的打击。

blow up 炸毁；充气：The soldiers blew up the bridge with gun powder. 战士们用炸药炸毁了这座桥。/I have to get my tires blown up. 我得给车胎打气了。

potential

a.   潜在的，可能的：A lot of potential buyers have shown special interest in the company. 很多潜在的购买者对这家公司特别感兴趣。/We are aware of the potential problems and have taken every precaution. 我们意识到潜在的问题，并已尽可能地采取了防范措施。

n.   潜力，潜能：The boy has great potential. 这个男孩有很大的潜力。

派   **potentially**

plain

n.   平原：The family crossed the plains in a covered wagon in 1860. 1860 年，那家人坐篷车穿过大平原。

a.   1. 朴素的；简单的：Her dress is quite plain. 她衣着很朴素。/The novel is plain in language. 这部小说的语言通俗易懂。

2. 清晰的，明白的：It's quite plain that they don't want to speak to us. 很明显他们不想跟我们说话。/Her voice is plain over the telephone. 电话里她的声音很清楚。

派   **plainly**

shoot

v.   1. 发射；射中；射击：The criminal shot a pistol at me. 罪犯用手枪向我射击。/The murderer was shot. 这个杀人犯被击毙了。

2. 射（门），投（篮）: He shot from the middle of the field and scored. 他从中场射门得分。

3. 飞快地移动，迅速上升: As we slowed down for the turn, a red car shot ahead of us. 正当我们要减速转弯时，一辆红色轿车迅速超过了我们。/Rents have shot up in the last few months. 在过去的几个月里，租金暴涨。

4. 发芽: The trees are shooting after the first rain of the spring. 第一场春雨过后，这些树发芽了。

5. 拍摄: This film was shot in Italy. 这部电影是在意大利拍的。

n. 1. 射击，发射: He made a shoot while driving. 他开车的时候打了一枪。

2. 嫩枝，幼苗: The young trees are putting out new shoots. 小树长出了新枝。

派 **shooting, shot**

---

worry

---

v. （使）烦恼，（使）发愁；（使）担心: Don't worry about the children; they are old enough to take care of themselves. 不要为孩子们担心，他们已长大，能照顾自己了。/ Don't worry me with such foolish questions. 不要拿这些愚蠢的问题来烦我。

n. 担心，发愁，忧虑: Too much worry has made her look like an old woman. 她烦恼太多，看上去像个老太婆。/Unemployment can be a cause of worry. 失业可能使人忧虑。

派 **worried**

Writing Practice

I. **Rewrite the following sentences, beginning with "*It....*"**

Example:

To drive a car without a license is illegal. ➔ It is illegal to drive a car without a license.

1. That she was not hurt in the accident was a miracle.

   _____

2. Where the package was sent from was far from clear.

   _____

3. To see a movie in the cinema is very enjoyable.

   _____

4. I was worried that she drove so fast.

_____

5. The man took a week to clean our garden.

_____

6. A lot of effort is needed to play the violin well.

_____

7. Bill appears likely to win the election.

_____

8. She seemed to have overused her budget.

_____

9. People agreed that no further changes would be made.

_____

10. His not taking the offer surprised everybody.

_____

11. Teaching small children requires a lot of patience.

_____

12. She found it impossible to do homework in a noisy room.

_____

II. **Rewrite the following sentences using the structure** *"it + be... that/what"*.

Example:

I'm most anxious about <u>the economics exam</u>. ➔ It is the economics exam that I am most anxious about.

1. Angela signed the check, not her husband.

_____

2. John arrived last Wednesday, not Thursday.

_____

3. I wrote to Peter, not to his brother.

_____

4. We met in Orleans, not Paris.

_____

5. (She's been seeing a doctor at Saint Mary's Hospital, but) she's having the operation in Johns Hopkins Medical Center.

_____

6. (They said they attended the convention because they liked volunteer work, but I think) they came to the event because they wanted free food.

_____

III. Rewrite the sentences according to the models.

Model A:

Original sentence: If you feel that is unreasonable, you are probably not a judge on the Illinois Appellate Court.

New sentence: If that strikes you as unreasonable, you're probably not a judge on the Illinois Appellate Court. (Text B)

1. It makes me feel strange that he never talks about his family.

_____

2. He always makes me feel that he is an honest man.

_____

Model B:

Original sentence: Chrysler should know that its cars will be used in bank robberies.

New sentence: Doesn't Chrysler know that its cars will be used in bank robberies? (Text B)

3. You slept enough last night.

_____

4. You can stay here for a few more days.

_____

Model C:

Original sentence: ... he was a member of the first group of military officers who willingly agreed to take an anthrax vaccine...

New sentence: ... he was among the first wave of military officers who willingly agreed to take an anthrax vaccine... (Text A)

5. A fifteen-year-old girl was one of the people who were injured.

_____

6. The new teacher was one of the speakers.

_____

Model D:

Original sentence: A short time after Timothy J. McVeigh blew up the Federal Building, the concept for an anti-terrorism center to prepare government officials, rescue workers and the public for another domestic attack was formulated.

New sentence: The concept for an anti-terrorism center to prepare government officials, rescue

workers and the public for another domestic attack was formulated shortly after Timothy J. McVeigh blew up the Federal Building.   (Text A)

7.   After you left for a short time, a man came into the office to look for you.

_____

8.   A short time before her death, she finished the novel.

_____

IV. Combine each set of the sentences into one, using the connective words or expressions provided.

1.   a. He spoke to Congress in the late 1990s about the potential dangers of unconventional warfare.

b. Shortly after that he willingly agreed to take an anthrax vaccine.

c. This was the first action of its kind.

d. He took the vaccine together with a group of military officers.

New sentence: _____ (and, among, who) (Text A)

2.   a. Something happened here.

b. We believe something.

c. Oklahoma City is a place.

d. You can bring serious people together.

e. You can talk about some complex issues.

f. The issues are associated with battling terrorism on U.S. soil.

New sentence: _____ (because of, where, and)   (Text A)

3.   a. Family members and survivors wanted an organization.

b. It looked to the future.

New sentence: _____ (that) (Text B)

4.   a. State leaders, survivors and relatives of the victims wanted a nonprofit organization.

b. It is part of the memorial complex.

c. The organization's broad mission is to prevent terrorism on U.S. soil.

New sentence: _____ (as, whose) (Text A)

5.   a. A three-judge panel of the court unanimously ruled.

b. Ceriale's family and the families of four other victims could sue the companies.

c. The companies produced guns.

d. The guns were used to kill their relatives.

New sentence: _____ (that) (Text B)

6. a. That certainly sounds bad.

 b. It's hard to understand something.

 c. Gun makers can be expected to stop their products from falling into the hands of criminals.

 New sentence: _____ (but, how)

 (Text B)

7. a. It's true.

 b. Gun makers must know it.

 c. Some of their products will be used to commit murder.

 New sentence: _____ (while, that)

 (Text B)

8. a. Suing them seems more ridiculous.

 b. It is ridiculous suing Smith & Wesson.

 c. Perhaps guns are widely seen as less legal than cars and baseball bats.

 New sentence: _____ (if, more...

 than, its because) (Text B)

V. Translate the following sentences into English.

1. 犯罪往往起源于贫穷。（outgrowth）
2. 制造商应该在设计阶段就发现潜在的问题。（potential）
3. 注意不要让这事再发生。（look to）
4. 这两件东西外表相同，但本质不同。（in essence）
5. 形势正发生变化，我们的计划必须相应地改变。（accordingly）
6. 这个工人开动了发动机。（set in motion）
7. 没有什么能阻止我们实现目标。（stop from）
8. 简单地说，公司需要削减开支。（briefly put）

VI. Translate the following paragraph into Chinese.

On May 14, hundreds of thousands of people gathered to march in Washington, as well as in 70 other cities around the country, to call for enactment of strong, sensible gun laws. For years, overwhelming majorities of Americans have supported such proposals as licensing of gun buyers and registration of gun sales, which will help keep guns out of the wrong hands and reduce gun violence. The National Rifle Association pretends not to be worried by the impact of the march, but it knows it is in trouble if a large number of those who participated remain active and are devoted to this issue.

VII. Practical English Writing

Directions: Guns as a kind of weapon can serve human beings in many ways. But they can also bring harm to human beings. You can first list some of the ways in which guns can be used for good purposes, and then some of the ways in which guns can serve bad people for evil intentions. Then discuss with your classmates and see what you can learn from each other. Finally, you should work out some measures to make sure that guns are used properly. During your discussion you can take some notes with the help of the following table.

Benefits Brought by Guns	Harm Brought by Guns	Measures to Control the Use of Guns
1.	1.	1.
2.	2.	2.
3.	3.	3.
4.	4.	4.
5.	5.	5.

Now summarize the result of your discussion.

My View on the Use and Control of Guns

7

Unit

Before Reading

Discuss the following questions in class.

1. Did you enjoy your childhood? What do you remember about it?
2. Can young children be trained into geniuses?

Reading

Make a Child into a Genius

1   Mozart and Thomas Edison are two of the many child prodigies[1] whom we encounter[2] in the history of geniuses. Who is a child prodigy? Can young children actually be trained to become mentally[3] superb[4]? Let's see.

2   Normal kids should always be allowed to enjoy their childhood. They go to school for roughly six hours a day, and the rest of the time would be distributed[5] to having fun and pursuing hobbies. Such is the typical[6] lifestyle of a young child.

3   Recently, however, young children have been trained in another way. Before the commencement[7] of formal education, young children were sent to nurseries[8] where they interacted with each other and were taught fundamentals[9] like the alphabet[10] by their teachers. Children who are supposed to be playing in their homes are now being flooded[11] by numerous[12] CD-ROMs[13] and other learning appliances[14]. More hours are spent at school, and fewer and fewer hours are spent at home. Is it thus[15] true that kids who are going to be trained into geniuses follow the routine lifestyle of an adult?

4   In order to supposedly[16] coach[17] future geniuses, the consumers' shelves[18] are being flooded with numerous new products supposedly to serve that noble[19] mission. Products such as baby Mozart tapes are rumored[20] to induce[21] spatial[22] intelligence in babies and other abilities related to music and the arts as well. To increase visual[23] recognition abilities, colorful books are also prepared. Furthermore, there are also products that are bilingual[24] seeking to train babies to learn more languages. Even microphones are said to be placed at the mother's belly[25] in order to enhance the fetus's[26] brain development. Are these all ways to produce a child prodigy?

5   Young children are often deprived[27] of their childhood by these genius-development methods. It is not really worth the effort trying so hard to convert[28] a child into a genius.

A child will not be happy if he has not had the opportunity to enjoy the carefree[29] years of his life — childhood. Only childhood can give a child the unique personality which will develop during those years, giving him many opportunities to explore new things in nature, learn from adults, as well as being able to interact with others. If the child really has a talent to show, the childhood years will allow him to display that talent.

6   Research has also shown that products inducing early intelligence development may do more harm than good. Children may be programmed towards learning certain things and their brain development or potential may be held back[30] as well. Thus, what is the solution here? Let children be children and enjoy their childhood years. Let those good times roll.

New Words and Expressions

1 prodigy ■ /'prɒdɪdʒɪ/ *n.* 天才，奇才（尤指神童）

2 encounter ▲ /ɪn'kaʊntə(r)/ *vt.* 遭遇，遇到

3 mentally # /'mentəlɪ/ *ad.* 智力上，脑力上

4 superb ▲ /sjʊ'pɜːb/ *a.* 极好的，杰出的

5 distribute /dɪ'strɪbjuːt/ *v.* 分配

6 typical /'tɪpɪkəl/ *a.* 典型的，有代表性的

7 commencement ■ /kə'mensmənt/ *n.* 开始；开端

8 nursery /'nɜːsərɪ/ *n.* 托儿所，保育室

9 fundamental /ˌfʌndə'mentəl/ *n.* [常 ~s] 基础

10 alphabet ★ /'ælfəbɪt/ *n.* （一种语言的）全部字母

11 flood /flʌd/ *v.* 淹没，涌入

12 numerous ★ /'njuːmərəs/ *a.* 众多的，许多的

13 CD-ROM *abbr.* [计]（信息容量极大的）光盘只读存储器

14 appliance ★ /ə'plaɪəns/ *n.* 工具，用具

15 thus /ðʌs/ *ad.* 因此，从而

16 supposedly # /sə'pəʊzɪdlɪ/ *ad.* 根据推测

17 coach /kəʊtʃ/ *vt.* 训练，指导

18 shelf /ʃelf/ *n.* 架子，搁板

19 noble /'nəʊbl/ *a.* 高尚的，崇高的

20 rumor ▲ /'ruːmə(r)/ *vt.* [常用被动语态] 谣传；传说

21 induce ★ /ɪn'djuːs/ *vt.* 引起，导致

22 spatial# /'speɪʃəl/ *a.* 空间的，与空间有关的

23 visual* /'vɪzjʊəl/ *a.* 视觉的，视力的

24 bilingual■ /baɪ'lɪŋwəl/ *a.* 使用两种语言的

25 belly■ /'belɪ/ *n.* 肚子，腹部

26 fetus■ /'fiːtəs/ *n.* 胎儿，胚胎

27 deprive■ /dɪ'praɪv/ *vt.* （*of*） 剥夺，使丧失

28 convert* /kən'vɜːt/ *vt.* （使）转变，（使）转化

29 carefree# /'keəfriː/ *a.* 无忧无虑的，轻松愉快的

30 hold back 阻碍

Proper Nouns

Mozart /'məʊtsɑːt/ 莫扎特 ［1756-1791，奥地利作曲家］

Thomas Edison /'tɒməs'edɪsən/ 托马斯·爱迪生 ［1847-1931，美国发明家］

 After Reading

A. Main Idea

- -

Complete the following diagram with the sentences or expressions given below.

1. Normal kids should always be allowed to enjoy their childhood.

2. flooded with numerous new products

3. Young children are often deprived of their childhood by these genius-development methods.

4. do more harm than good

5. Let children be children and enjoy their childhood years.

6. young children were sent to nurseries

7. numerous CD-ROMs and other learning appliances

8. More hours are spent at school, and fewer and fewer hours are spent at home.

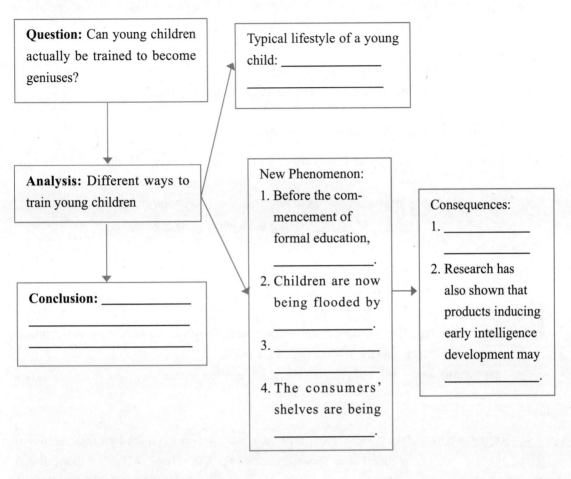

Question: Can young children actually be trained to become geniuses?

Typical lifestyle of a young child: _____
_____

Analysis: Different ways to train young children

New Phenomenon:
1. Before the commencement of formal education, _____.
2. Children are now being flooded by _____.
3. _____
_____
4. The consumers' shelves are being _____.

Consequences:
1. _____
_____
2. Research has also shown that products inducing early intelligence development may _____.

Conclusion: _____
_____
_____

B. Detailed Understanding

I. Tell if the following statements are true (T) or false (F) according to the text.

1. _____ According to the author, young children can be trained to become mentally superb.

2. _____ A child will be happy if he has had the opportunity to enjoy the carefree years of his life—childhood.

3. _____ Numerous CD-ROMs and other learning appliances have benefited children a lot.

4. _____ Before the commencement of formal education, young children were sent to nurseries where they had fun and pursued hobbies.

5. _____ Children now spend fewer and fewer hours at home.

6. _____ The genius-development methods are effective in making a child into a genius.

7. _____ The author thinks it is really worth the effort trying hard to convert a child into a genius.

8. _____ Baby Mozart tapes can induce spatial intelligence in babies and other abilities related to music and the arts as well.

9. _____ Research has shown that products inducing early intelligence development have probably done more harm than good.

10. _____ Children should be allowed to enjoy their childhood.

II. Explain the sentences by filling in the blanks.

1. **Text sentence:** They go to school for roughly six hours a day, and the rest of the time would be distributed to having fun and pursuing hobbies.

   Interpretation: After studying at school for about six hours a day, children would _____ having fun and pursuing hobbies.

2. **Text sentence:** Such is the typical lifestyle of a young child.

   Interpretation: A kid usually _____.

3. **Text sentence:** ... the consumers' shelves are being flooded with numerous new products supposedly to serve that noble mission.

   Interpretation: ... the consumers' shelves are filled with many new products which _____.

4. **Text sentence:** Products such as baby Mozart tapes are rumored to induce spatial intelligence in babies and other abilities related to music and the arts as well.

   Interpretation: It is said that _____ induce both _____ and other abilities that _____.

5. **Text sentence:** Furthermore, there are also products that are bilingual seeking to train babies to learn more languages.

   Interpretation: In addition, there are also products that use _____.

6. **Text sentence:** A child will not be happy if he has not had the opportunity to enjoy the carefree years of his life—childhood.

   Interpretation: A child will not be happy unless _____ his childhood and _____.

7. **Text sentence:** Research has also shown that products inducing early intelligence development may do more harm than good.

Interpretation: Research has also shown that products that _____ may have _____ rather than _____ effects.

8. **Text sentence:** Children may be programmed towards learning certain things and their brain development or potential may be held back as well.

   Interpretation: Children may be taught in a special way to learn certain things but this may _____.

C. Detailed Study of the Text

1 **Mozart and Thomas Edison are two of the many child prodigies whom we encounter in the history of geniuses.** (Para. 1) 在天才人物历史上出现过许多神童,莫扎特和托马斯·爱迪生就是其中的两位。

Mozart: 莫扎特 (1756-1791),奥地利作曲家,维也纳古典乐派主要代表,五岁开始作曲,写出大量作品,主要有歌剧《费加罗的婚礼》、《唐璜》、《魔笛》及交响曲、协奏曲、室内乐等。

Thomas Edison: 托马斯·爱迪生 (1847-1931),美国发明家,拥有白炽电灯、留声机、炭精话筒、口述录音机、电影放映机、蓄电池等 1,093 项发明专利。幼时对周围事物怀有极大好奇心,极爱发问,课堂上也不例外,因而被教师用皮带抽打,母亲不忍儿子被虐待,不再让他上学。因此,他一生所受的正式教育也仅有三个月。后来由母亲启蒙,自学成材。

2 **They go to school for roughly six hours a day, and the rest of the time would be distributed to having fun and pursuing hobbies.** (Para. 2) 他们每天大约六个小时上学,其余的时间玩,做点自己喜欢的事。

be distributed to 表示"被分配",其中 to 为介词,后面跟动作时,应用动词的 -ing 形式。

3 **Such is the typical lifestyle of a young child.** (Para. 2) 这就是一个儿童典型的生活方式。

such 作代词时多作主语,指上面说的情况,意思为"这(些)"。例如:Such is life. 这就是生活。/Such are the results. 结果就是这样。

4 **It is not really worth the effort trying so hard to convert a child into a genius.** (Para. 5) 花这么大的力气去把孩子变成天才实在是得不偿失。

句中的 it 作形式主语,真正的主语是 trying so hard to convert a child into a genius,由于其较长,所以由 it 替代,以避免句子头重脚轻。

not worth the effort 意为"不上算,得不偿失;不值得"。

5 **Only childhood can give a child the unique personality which will develop during those years, giving him many opportunities to explore new things in nature, learn from adults, as well as being able to interact with others.** (Para. 5) 唯有童年才能发展孩子独特的个性,才能给孩子许多机会去探索自然界中的新鲜事物,去向大人们学习,并能与人交往。

句中的 which will develop during those years 是定语从句，修饰 the unique personality。giving him... 为现在分词短语，进一步修饰 childhood。

6 **Children may be programmed towards learning certain things and their brain development or potential may be held back as well.** (Para. 6) 儿童可以通过特定的程序学会某些东西，但他们的大脑发育或潜能也许会因此受到抑制。

短语 hold back 在句中的意思是"阻碍（发展）"，如：You could become a good musician, but your lack of practice is holding you back. 你有可能成为一名优秀的音乐家，但是练习不足正在妨碍你的发展。此短语也有"控制（感情、眼泪等）"的意思，如：Jim was able to hold back his anger and avoided a fight. 吉姆强忍住怒火，避免了一场争斗。

D. Talking About the Text

Work in pairs. Ask and answer the following questions first and then put your answers together to make an oral composition.

1. What should normal kids be allowed to do?
2. What would normal kids do after school?
3. What are young children asked to do before the commencement of formal education?
4. What are children flooded by?
5. Why have young children been trained in this way?
6. What is the noble mission that numerous new products are supposed to serve?
7. What does research show about the effects of such products?
8. Will the child be happy in this way?
9. Why is childhood important in the development of one's character?
10. What is the conclusion made by the author?

E. Vocabulary Practice

I. Fill in the blanks with the new words or expressions from Text A.

1. We hope the new machine will work faster, _____ reducing our costs.
2. The office was _____ with complaints.
3. She felt that she was _____ from further promotion.
4. _____ speaking, I would say that about 100 people attended the exhibition.
5. Tom is a _____ absent-minded professor.
6. She was _____ of her sight by the accident.

7. The author by his choice of words has _____ a particular frame of mind in the reader.

8. Household electric _____ made in China are becoming more popular around the world.

9. We'll soon _____ him to our way of thinking.

10. Leaving university brought my _____ days to an end.

II. Complete the following dialogs with appropriate words or expressions from Text A.

1. A: Hey, where did you find the journal? I need it, too.

   B: Right here on the _____. Don't worry, John. We can share it.

2. A: I spend my leisure time playing tennis.

   B: That's a _____ way to stay in shape and also have a good time.

3. A: I like their new advertisement.

   B: It's a _____ example of their marketing strategy.

4. A: How did you feel about your travel?

   B: It was amazing! You have absolutely no idea how many interesting people and stories I have _____!

5. A: Where are you from originally?

   B: I'm from Washington, D.C. I think it's the most _____ city in the United States.

6. A: It is _____ that the bill will get passed in Congress.

   B: That would be wonderful.

7. A: _____ the company offered you a pay rise of 50%, would you be so determined to leave and look for a job elsewhere?

   B: Yes, I've set my mind on it. I'd like to find a job that can better show my ability.

8. A: Do I have a beer _____?

   B: I'm afraid so.

9. A: Despite _____ failures, they continued with their experiment.

   B: And they finally succeeded.

10. A: What's the purpose of the training program?

    B: To help some handicapped people recover physically and _____.

Text B

 Before Reading

Discuss the following questions in class.

1. Did your parents often talk with you when you were a child?
2. Do you think communication is vital to children? Why or why not?

 Reading

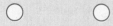 

A Hidden Blessing[1]

1　　Everyone agrees that communication is important—with your husband or wife, your co-workers, your boss. When it comes to children, however, communication is vital[2], and unfortunately, often ignored. Every parent knows that children need attention, regardless of age. When children are ignored, they can develop a wide range of deficiencies[3] ranging from a lack of social skills and moral values to poor self-image, depression and poor academic performance. Open communication with adults is especially essential for teenagers as they learn to define[4] their places in the world.

2　　Research clearly shows that one-on-one interaction with children is foundational[5] to their moral[6] development, as well as for the development of their critical[7] thinking skills and maturity[8]. Children gain skills and learn about successful living and relationship building through their interactions with parents, teachers and other significant[9] adults who enter their lives.

3　　Experts say that what most powerfully impacts[10] a child may seem unimportant at first. Random[11] acts of caring and natural interest are far more influential[12] than planned activities and teaching, though these are also necessary. Dr. Steven Glenn and Jane Nelsen who authored *Raising Self-Reliant*[13] *Children* explain how dialogue develops a child's self-image. When young people are given the opportunity to engage in meaningful dialogue with people they admire, they begin to perceive[14] themselves as significant in the sight of adults. In addition, they absorb[15] our values, thinking patterns and ideals.

4　　What are some of the moments that make a difference in a child's life? When we praise them for what they do well... When we teach and encourage them... When we are sensitive[16] to their needs and feelings... When we give them the support they need... When we are good role models. The opportunities are endless to initiate[17] a conversation with a child and give him/her your undivided attention.

5 　　Most parents do not spend enough time talking with their children. Studies show quality time ranges an average of 7 to 15 minutes a day. Television has displaced[18] family interaction in the home. Children generally spend hours a day in front of the TV set, both with and without their parents. The television culture may have harmful consequences[19] as it takes over dialogue which is fundamental for children to develop critical thinking, moral understanding, bonding[20] and trust. Research is also showing that too much television watching has negative[21] effects on brain development and learning.

6 　　We have all heard the African proverb: "It takes an entire village to raise a child." As extended families disappear, children need interaction with other significant adults more than ever. Take a moment and think about your own childhood and about the grown-ups in your life, other than your parents, who positively[22] influenced you and guided you. What were their outstanding[23] qualities that made an impression on you? Were they accepting, sensitive, patient, genuine or fun? The interactions we had with significant adults changed our lives. Now is our turn to play this role for children around us, building relationships that will nourish[24] their souls and develop solid foundations in their lives. All children appreciate[25] attention because it conveys[26] to them that they are loved. Your gift can be as simple as offering an encouraging word, showing kindness or comforting a child whose feelings are hurt.

7 　　As we give to children, we also reap[27] a reward. As we put our busy lives on hold for a few moments to bond with a child, we also benefit from the exchange of love, warmth and sincerity[28]. Talking to a child brings a ray[29] of sunshine to our lives. Let us not miss[30] these priceless[31] opportunities to make a difference in a child's life and receive a wonderful blessing.

New Words and Expressions

1　bless ■ /bles/ *vt.* 为……祈神赐福（或保佑）
　　blessing ■ /'blesɪŋ/ *n.* 祝福
2　vital /'vaɪtəl/ *a.* 非常必需的，极其重要的
3　deficiency ■ /dɪ'fɪʃənsɪ/ *n.* 缺点，缺陷
4　define /dɪ'faɪn/ *v.* 明确

5　foundational# /faʊnˈdeɪʃənəl/ a. 基本的，基础的

6　moral★ /ˈmɒrəl/ a. 道德（上）的

7　critical /ˈkrɪtɪkəl/ a. 批评的，批判的

8　maturity# /məˈtjʊərətɪ/ n. 成熟的状态或品质

9　significant /sɪgˈnɪfɪkənt/ a. 影响深远的

10　impact /ɪmˈpækt/ v. 对……有影响

11　random★ /ˈrændəm/ a. 任意的，随机的

12　influential▲ /ˌɪnfluˈenʃəl/ a. 有影响的

13　self-reliant /ˌselfrɪˈlaɪənt/ a. 依靠自己的，自力更生的

14　perceive /pəˈsiːv/ vt. 意识到，认识到

15　absorb★ /əbˈsɔːb/ vt. 吸收

16　sensitive /ˈsensɪtɪv/ a. 体贴入微的

17　initiate■ /ɪˈnɪʃɪət/ vt. 开始

18　displace■ /dɪsˈpleɪs/ vt. 取代，替代

19　consequences /ˈkɒnsɪkwəns/ n. 后果

20　bond★ /bɒnd/ v. 形成亲密的人际关系

　　bonding★ /ˈbɒndɪŋ/ n. 人与人之间亲密关系的形成

21　negative /ˈnegətɪv/ a. 消极的

22　positively /ˈpɒzətɪvlɪ/ ad. 积极地

23　outstanding /ˌaʊtˈstændɪŋ/ a. 突出的，杰出的

24　nourish▲ /ˈnʌrɪʃ/ vt. 滋养，养育，培养

25　appreciate /əˈpriːʃɪeɪt/ v. 欣赏，赏识

26　convey /kənˈveɪ/ v. 表达（感情等）

27　reap■ /riːp/ vt. 获得，得到

28　sincerity# /sɪnˈserɪtɪ/ n. 真诚，诚挚

29　ray /reɪ/ n. 光线

30　miss /mɪs/ vt. 错过；失去

31　priceless# /ˈpraɪslɪs/ a. 珍贵的，无价的

Proper Nouns

Steven Glenn /ˈstiːvənglen/ 斯蒂文·格伦 [人名]

Jane Nelsen /ˈdʒeɪnˈnelsən/ 简·纳尔逊 [人名]

After Reading

A. Main Idea

Complete the following diagram with the expressions given below.

1. One-on-one interaction with children
2. Parents' failure to spend enough time talking with their children
3. Open communication with adults
4. need for interactions
5. develop a wide range of deficiencies
6. Both children and parents
7. What most powerfully impacts a child
8. to give a child undivided attention

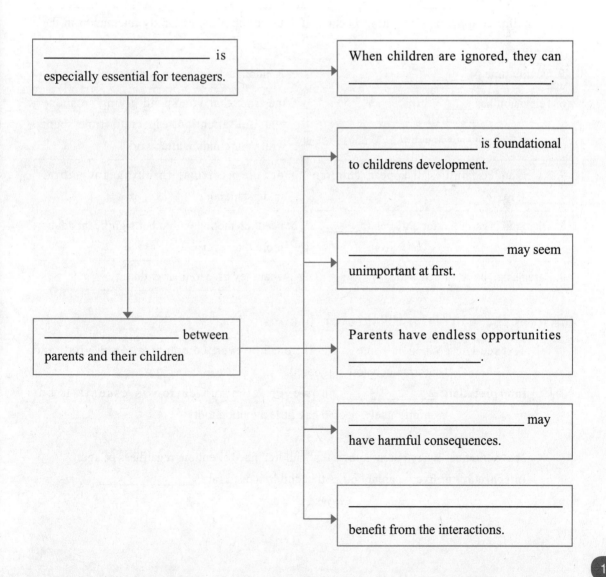

_____ is especially essential for teenagers.

When children are ignored, they can _____.

_____ is foundational to childrens development.

_____ may seem unimportant at first.

_____ between parents and their children

Parents have endless opportunities _____.

_____ may have harmful consequences.

_____ benefit from the interactions.

B. Detailed Understanding

I. Make correct statements according to the text by combining appropriate sentence parts in Column A with those in Column B.

Column A	Column B
1. _____ are far more influential than planned activities and teaching.	a. their interactions with parents, teachers and other significant adults who enter their lives
2. When children talk with people they admire, they _____.	b. think they are considered important
3. Children gain skills and learn about successful living and relationship building through _____.	c. it conveys to them that they are loved
4. All children appreciate attention because _____.	d. taken the place of family interaction in the home
5. Quality time is _____.	e. Random acts of caring and natural interest
6. Television has _____.	f. the time that you spend giving someone your full attention, especially time spent with your children after work
7. As extended families disappear, children _____.	g. develop critical thinking and moral understanding
8. Dialogue is essential for children to _____.	h. need interaction with other significant adults more than ever
9. Parents can be rewarded with _____.	i. what they give to their children

II. Explain the sentences by filling in the blanks.

1. **Text sentence:** When it comes to children, however, communication is vital, and unfortunately, often ignored.

   Interpretation: _____, however, communication is essential, but unfortunately, people pay little attention to it.

2. **Text sentence:** Every parent knows that children need attention, regardless of age.

   Interpretation: Every parent knows that children need care, _____.

3. **Text sentence:** Research clearly shows that one-on-one interaction with children is foundational to their moral development, as well as for the development of their critical thinking skills and maturity.

    Interpretation: Research clearly shows that _____ not only to developing their abilities of critical thinking but also to making them more mature.

4. **Text sentence:** Experts say that what most powerfully impacts a child may seem unimportant at first.

    Interpretation: According to experts, _____.

5. **Text sentence:** Random acts of caring and natural interest are far more influential than planned activities and teaching...

    Interpretation: Unplanned caring and natural interest have greater influence _____...

6. **Text sentence:** In addition, they absorb our values, thinking patterns and ideals.

    Interpretation: _____ our values, thinking patterns and ideals.

7. **Text sentence:** The opportunities are endless to initiate a conversation with a child and give him/her your undivided attention.

    Interpretation: There are _____ and give him/her your full attention.

8. **Text sentence:** It takes an entire village to raise a child.

    Interpretation: Not only the parents but also _____.

9. **Text sentence:** Now is our turn to play this role for children around us.

    Interpretation: Now we should _____.

C. Detailed Study of the Text

----

1 **When it comes to children, however, communication is vital, and unfortunately, often ignored.** (Para. 1) 而对孩子来说，沟通更为至关重要，可惜，这一点常常被忽略。

   句式 when it comes to sth./doing sth. 表示"就……而言"、"谈到……"，用于引出一个论点或话题，可以置于句首也可以置于句尾。如：When it comes to politics, I know nothing. 关于政治，我一窍不通。

2 **When we are good role models.** (Para. 4) 当我们以身作则的时候。

   这里的 role model 是"行为榜样"，在某一特定的行为或社会角色中供他人模仿或作为榜样的人。

3　Studies show quality time ranges an average of 7 to 15 minutes a day. (Para. 5) 有研究表明，父母与子女共同活动的时间每天平均为 7 至 15 分钟。

quality time 有效时间，本文中指父母与子女共同活动的时间。

4　As extended families disappear, children need interaction with other significant adults more than ever. (Para. 6) 随着大家庭的消失，孩子们比以往任何时候都更需要与其他重要的成人交往。

extended family 大家庭，扩大的家庭（如与祖父母、已婚子女等共居的数代同堂家庭），与之相对的为 nuclear family 小家庭（仅由夫妻与子女组成）。

5　Take a moment and think about your own childhood and about the grown-ups in your life, other than your parents, who positively influenced you and guided you. (Para. 6) 坐下来想一想你的童年，想一想在你的一生中除你父母之外对你产生过积极影响并指导过你的成年人。

这里 other than your parents 是插入语。other than 此处意思为"除了……，除……之外"，如：There is nobody here other than me. 这里除了我没别人。 另外，句中 who positively influenced you and guided you 是个限制性定语从句，修饰 the grown-ups。

6　As we put our busy lives on hold for a few moments to bond with a child, we also benefit from the exchange of love, warmth and sincerity. (Para. 7) 当我们把繁忙的生活搁置片刻与孩子形成亲密关系的时候，在爱、温暖与真诚的交流中，我们也会受益。

put something on hold 意为"推迟"；benefit from 意为"从……中受益，获益"。

D. Further Work on the Text

Write down at least three more comprehension questions of your own. Work in pairs and ask each other these questions. If you can't answer any of these questions, ask your classmates or the teacher for help.

1. _____
2. _____
3. _____

E. Vocabulary Practice

I. Find the word that does NOT belong to each group.

1. A. husband   B. co-worker   C. wife   D. child
2. A. vital   B. essential   C. significant   D. unimportant
3. A. development   B. communication   C. interaction   D. solution
4. A. priceless   B. valueless   C. invaluable   D. valuable
5. A. random   B. grown-up   C. moral   D. self-reliant

II. Complete the following sentences with appropriate words in their correct form.

1. **admire, admiration, admirer**
   1) He's always looking in the mirror, _____ himself.
   2) She has many _____.
   3) His new car made him the _____ of his friends.

2. **extend, extension, extensive, extended**
   1) I planned an _____ of my holiday.
   2) My younger brother had an _____ vacation in the Alps.
   3) The storm caused _____ damage.
   4) We will eventually _____ the road to the power station.

3. **mature, maturity, maturely, immature**
   1) He is in his full _____ as a Shakespearean actor.
   2) Wine and judgment _____ with age.
   3) But he turned out to be _____ and irresponsible.
   4) She acts very _____ for her age.

4. **critic, critical, criticize, criticism**
   1) It takes years to develop one's _____ ability.
   2) Any _____ you can make on my draft will be greatly appreciated.
   3) He is his own most severe _____.
   4) She's always _____ her husband for being slow.

5. **sincere, sincerity, sincerely, insincere**
   1) Complete _____ can affect even metal and stone.
   2) May everything beautiful be condensed into this card. I _____ wish you happiness, cheerfulness and success.
   3) The most exhausting thing in life is being _____.
   4) A friend is a person with whom I may be _____. Before him, I may think aloud.

Word Study

flood

n.   1. 洪水；水灾：The flood swept away many houses. 洪水冲走了许多房屋。

      2. 大量，滔滔不绝：a flood of tears 泪如泉涌 /a flood of people 人潮涌动 /floods of rain 倾盆大雨

v.   1. 淹没；使泛滥：Every spring the river floods the valley. 每年春天河水都在山谷泛滥。

      2. 溢满；覆满：The room was flooded with sunlight. 房间里充满了阳光。/In the past few decades foreign goods flooded the markets of the developing countries. 过去几十年中，外国货充斥着发展中国家的市场。

      flood in 大量涌来，涌到；冲进：Applications flooded in. 申请书大量涌来。

essential

a.   1. (*to*) 必要的，必不可少的：Food is essential to life. 食物是维持生命不可或缺的。

      2. 本质的；实质的；基本的：Love of fair play is said to be an essential part of the English character. 喜爱公平竞争据说是英国人性格中的重要方面。

n.   [常 *pl.*] 本质；要素；要点：In considering a problem, one must grasp its essentials. 观察问题要抓住它的本质。/the essentials of grammar 语法要点

派   **essentially**

roll

v.   1. 滚动；滚落；翻落：The ball rolled into the hole. 球滚进了洞里。

      2. （使）摇摆；（使）摇晃：Heavy seas rolled the ship. 汹涌的海浪使船左右摇晃。

      3. 碾；轧：She rolled out the flour and water mixture to make bread. 她把面粉和水和成的面团擀好来做面包。

      4. 卷；绕：He rolled up the map. 他把地图卷起来。

n.   1. 一卷；卷状物：a roll of film 一卷胶卷

      2. 面包卷：a roll of bread 一个面包卷儿

      3. 名单；名册：the college roll 大学名册 /call the roll 点名

      roll in 滚滚而来，大量涌来：Offers of help are rolling in. 援助源源而来。

      roll out   1. 展开：We rolled out the red carpet for the important visitor. 我们铺红地毯欢迎这位重要来宾。

                      2. 压平；辗平：roll out the pastry 把面团擀开

      roll up   1. 卷起（袖子或裤管）：He rolled up his sleeves. 他卷起双袖。

2. 出现；来到；姗姗来迟：He rolled up last. 他最后一个到。

suppose

v.   1. 料想；猜想；认为：I suppose he has missed the train. 我想他没赶上火车。/I suppose it will rain. 我认为天会下雨。

2. （用于祈使句）让，假设：Suppose we dine together. 我们一块儿吃饭吧。

be supposed to   1. 有义务做……，应该：You are not supposed to smoke here. 你不应该在这里抽烟。

2. 意图是，打算，旨在：This law is supposed to help the poor. 这条法律旨在帮助穷人。

派   supposed, supposedly, supposition

value

n.   1. 价值：What is the value of your house? 你的房子值多少钱？

2. 有用性；重要性：the value of education 教育的重要性

3. [*pl.*] 价值观念，标准：ethical values 伦理标准

v.   1. 尊重；重视；珍视：I have always valued his advice. 我一向尊重他的意见。

2. 评价；估价：I valued the house at 500,000 *yuan*. 我估计这所房子值 50 万元。

派   valuable, invaluable, valueless

Writing Practice

I.   **Underline the connecting words in the following paragraph, paying particular attention to pronouns, articles, adverbs and conjunctions.**

Madrid is the capital of Spain. Phillip II moved his court to Madrid in 1561 and made Madrid the only court. With this decision, he passed over many older cities in Spain. These cities, like Seville, Cordoba, Zaragoza, Toledo, had felt they had a greater claim. Madrid was the newest city in Spain. Today, it is the largest city in Spain. The king's palace is there and so are the Parliament and the main government offices. Therefore, Madrid is now the political, cultural and financial center of Spain.

II. Rewrite the following sentences with the words provided in brackets.

1. My neighbors have got a new cat. The cat makes an awful noise at night. (it)

   ___

2. People travel all over the world. There is still a lot of cultural misunderstanding. (although)

   ___

3. Give them the money. They earned the money in a hard way. (which)

   ___

4. Some people can never forget the horrors of war. This is the most difficult thing after a war. (which)

   ___

5. In less than ten years, there will be only six car manufacturers in the world. The six manufacturers will be the ones in the United States, Japan and Germany. (they)

   ___

6. I got up early. I had to get on the first train to San Francisco. (because)

   ___

7. It was a very violent crime. The criminal must stay in jail for life. (so)

   ___

8. I know it's dangerous. I want to go to Iraq. (nevertheless)

   ___

9. I've never liked my neighbor. I've spent a lot of time helping them. (however)

   ___

10. He had 7 houses and 2 farms. He never worked in his life. (and)

    ___

III. Choose the correct connective.

1. Some of his photographs had won prizes in competitions. *So that/Consequently*, he thought of himself as a professional photographer.
2. The graphics in that new computer game are quite good. *Even so/Even though*, I got bored with it very soon.
3. He was refused entry to the country. *Though/Instead* he was forced to return to Spain.
4. John has lived in the village for 20 years. *Even though/Nevertheless* the locals still consider him an outsider.
5. They met for tea at a café in High Street and *afterwards/since* they went shopping.

IV. Rewrite the sentences according to the models.

Model A:

Original sentence: Products like baby Mozart tapes are rumored to induce spatial intelligence in babies...

New sentence: Products such as baby Mozart tapes are rumored to induce spatial intelligence in babies... (Text A)

1. The local community still relies on traditional industries like farming and mining.

   _____

2. A boy like John can never be expected to lose.

   _____

Model B:

Original sentence: A child will not be happy unless he has had the opportunity to enjoy the carefree years of his life—childhood.

New sentence: A child will not be happy if he has not had the opportunity to enjoy the carefree years of his life—childhood. (Text A)

3. I will not forgive you unless you give me your apology in person.

   _____

4. This kind of animal will disappear from the earth soon unless protective measures are taken immediately.

   _____

Model C:

Original sentence: ... the thing which most powerfully impacts a child may seem unimportant at first.

New sentence: ... what most powerfully impacts a child may seem unimportant at first. (Text B)

5. The thing which you want has been sent here.

   _____

6. The thing which he needs to do is more practice.

   _____

Model D:

Original sentence: We had interactions with significant adults and these interactions changed our lives.

New sentence: The interactions we had with significant adults changed our lives. (Text B)

7. The audience showed positive reactions and these reactions were very encouraging to the film director.

   _____

8. His friend brought him the medicine and this medicine saved his life.

   _____

V. **Combine each set of the sentences into one, using the connective words or expressions provided.**

1. a. Before the commencement of formal education, young children were sent to nurseries.

    b. They interacted with each other.

    c. They were taught fundamentals like the alphabet by their teachers.

    New sentence: _____ (where, and)

    (Text A)

2. a. Children may be programmed towards learning certain things.

    b. Their brain development or potential may be held back.

    New sentence: _____ (and, as well)

    (Text A)

3. a. Children are supposed to be playing in their homes.

    b. Children are now being flooded by numerous CD-ROMs.

    c. They are being flooded by other learning appliances.

    New sentence: _____ (who, and)

    (Text A)

4. a. Open communication with adults is especially essential for teenagers.

    b. Teenagers learn to define their place in the world.

    New sentence: _____ (as) (Text B)

5. a. The television culture may have harmful consequences.

    b. The television culture takes over dialogue.

    c. The dialogue is fundamental for children to develop critical thinking, moral understanding, bonding and trust.

    New sentence: _____ (as, which)

    (Text B)

6. a. All children appreciate attention.

    b. Attention conveys to them that they are loved.

    New sentence: _____ (because)

    (Text B)

7. a. Planned activities and teaching are influential.

    b. Random acts of caring are far more influential.

    c. Natural interest is far more influential.

    d. Planned activities and teaching are also necessary.

    New sentence: _____ (than, and, though) (Text B)

8. a. Young people are given the opportunity to engage in meaningful dialog with certain people.

b. They admire these people.

c. They begin to perceive themselves as significant in the eyes of adults.

New sentence: _____ (when) (Text B)

VI. Translate the following sentences into English.

1. 乔在体育方面不行，但在数学方面他是班上最好的。(come to)
2. 当你家里有了孩子时，你就有干不完的活。(endless)
3. 船应该在什么时候开? (be supposed to)
4. 我做这个是为了帮助他。(benefit)
5. 他是个有批判眼光的人。(critical)
6. 你今天或明天去没有多大关系。（make... difference）
7. 他感到他不会得到进一步的提升。(hold back, promotion)
8. 如果你给他留言，我会转告他的。(convey)

VII. Translate the following paragraph into Chinese.

It is a good idea for parents to monitor (监控) the amount as well as the kind of television their preschool child watches. If your child appears to be crazy about war plays and weapons, you'd better control his viewing. Controlling viewing is easier to do during the preschool years than during the school years, so you should initiate a pattern of restricted television watching now.

VIII. Practical English Writing

Directions: Now that you are at college, some people may say that your parents have been very successful in raising you. But looking back on the way your parents raised you, you may find certain things that your parents should have done but they didn't do, or things that they shouldn't have done. Now suppose you are going to have a child of your own. What would you do or not do while raising your child? You should also give reasons for your decisions. The following table will help you take some notes while you discuss the question with your classmates.

Things You Would Do	**Things You Would Not Do**	**Reasons for Your Decisions**
1	1	1
2	2	2
3	3	3
4	4	4
5	5	5

Basing on what you have talked about in your discussion, write a short passage on your view of child education.

If I Had a Child

_____

_____

_____

_____

_____

_____

_____

8
Unit

Text A

 Before Reading

Discuss the following questions in class.

1. Name one of the world heritage sites in China and give a description of it.
2. What do you think are the effective ways to protect world heritage sites?

 Reading

Sichuan Takes Pains to Protect World Heritage[1] Sites

1   The world's tallest Buddha statue in Leshan City, in southwest China's Sichuan Province, is undergoing[2] its second round of facelift[3] since April.

2   The project aims to make the statue, which is more than 1,200 years old, weatherproof[4], by installing drainage[5] devices and protecting the statue's feet against pounding[6] waves.

3   The facelift project is the first one in China to use a World Bank loan to repair and restore[7] a world heritage site, according to Zeng Zhiliang, an engineer in the field of ancient relics[8] and architecture[9], who heads[10] a 20-strong team to do the work.

4   The United Nations Educational, Scientific and Cultural Organization (UNESCO) will investigate into the protection[11] of natural and cultural heritage sites worldwide this year.

5   The giant Leshan statue, which was included on the World Cultural Heritage list in 1996, has suffered apparent damage from the wind, water, acid[12] rains, and wear and tear[13] by visitors for years. China input 250 million *yuan* (US$30 million) to "beautify"[14] the giant statue from March 2001 to the end of last year.

6   The Buddha statue, sitting squarely[15] on a cliff[16], 71 meters from top to bottom and 28 meters across, is 18 meters higher than the standing Buddha statue in Bamian Valley, Afghanistan, once thought to be the tallest in the world.

7   China boasts[17] a total of 28 world natural and cultural heritage sites. Four are in southwestern Sichuan Province, second only to the capital Beijing. All world heritage sites in Sichuan are well preserved and in good condition.

8   The province has put into effect a regulation[18] specially drafted[19] for protecting local world heritage sites, the first local law of its kind enacted[20] in China and one of the few similar regulations in the world.

9   Sichuan's regulation bans the construction of hotels, guest houses, development zones[21] and other related facilities[22] in the core area of a world heritage site, calling for enhanced efforts to protect ecological[23] environment and the wide use of environment-friendly vehicles with clean fuel[24] in and around world heritage sites. The regulation also strictly specifies[25] the discharge[26] of polluted water, smoke and dust as well as the treatment of domestic waste.

10   With this regulation in effect, Sichuan Province is aiming to preserve its world heritage sites for future generations, a provincial[27] official said.

New Words and Expressions

1   heritage ■ /'herɪtɪdʒ/ n. 遗产
2   undergo ▲ /ˌʌndə'gəʊ/ vt. 经历，遭受
3   facelift ■ /'feɪslɪft/ n. （建筑物等的）翻新，整修
4   weatherproof # /'weðəpruːf/ a. 防风雨的，不受气候影响的
5   drainage ■ /'dreɪnɪdʒ/ n. 排水，放水
6   pounding /'paʊndɪŋ/ n. 猛击声
7   restore ▲ /rɪ'stɔː(r)/ vt. 修复，整修
8   relic ■ /'relɪk/ n. 遗迹，遗物，遗俗
9   architecture ▲ /'ɑːkɪtektʃə(r)/ n. 建筑学，建筑业
10   head /hed/ v. 带领
11   protection # /prəʊ'tekʃən/ n. 保护，防护
12   acid ▲ /'æsɪd/ a. 酸性的
13   wear and tear 消磨，消耗，磨损
14   beautify # /'bjuːtɪfaɪ/ vt. 美化，装饰
15   squarely # /'skweəlɪ/ ad. 正好，不偏不倚地
16   cliff ▲ /klɪf/ n. （尤指海边的）悬崖，峭壁
17   boast /bəʊst/ v. （贬）夸口
18   regulation ★ /ˌregjʊ'leɪʃən/ n. 规章，法规
19   draft ▲ /drɑːft/ vt. 起草，草拟
20   enact ■ /ɪ'nækt/ vt. 制定（法律）；通过（法案等）
21   zone ★ /zəʊn/ n. 地区，区域

22 facility /fəˈsɪlətɪ/ n. [pl.] 设施，设备
23 ecological# /ˌiːkəˈlɒdʒɪkəl/ a. 生态的，生态学的
24 fuel* /fjʊəl/ n. 燃料，燃烧剂
25 specify▲ /ˈspesɪfaɪ/ vt. 明确说明，具体指定
26 discharge* /dɪsˈtʃɑːdʒ/ n. 排放，放出
27 provincial# /prəʊˈvɪnʃəl/ a. 省级行政区的

Proper Nouns

Bamian Valley 巴米扬河谷
Afghanistan /æfˈgænɪstæn/ 阿富汗
Buddha /ˈbʊdə/ 佛陀 [佛教徒对释迦牟尼的尊称]
UNESCO /juːˈneskəʊ/ 联合国教育、科学及文化组织 [简称联合国教科文组织]

 After Reading

A. Main Idea

Complete the following diagram with the expressions given below.

1. the repair and restoring projects of the giant Leshan Statue

2. calling for the wide use of environment-friendly vehicles with clean fuel in and around world heritage sites

3. the facelift project since April

4. banning the construction of hotels, guest houses, development zones and other related facilities in the core area of a world heritage site

5. enacting a regulation specially drafted for protecting world heritage sites

6. strictly specifying the discharge of polluted water, smoke and dust as well as the treatment of domestic waste

7. beautifying the statue from March 2001 to the end of last year

```
┌─────────────────────────────────────────────────────┐
│   Sichuan Takes Pains to Protect World Heritage Sites │
└─────────────────────────────────────────────────────┘
```

┌──────────────────────────┐   ┌──────────────────────────┐
│ Fact 1: _____ │   │ Fact 2: _____ │
│ _____ │   │ _____ │
└──────────────────────────┘   └──────────────────────────┘

┌────────────────────┐ ┌────────────────────┐ ┌──────────────────────────┐
│ Sponsored by World │ │ Funded by the Chinese│ │ Specific Measures: │
│ Bank: │ │ Government: │ │ 1. _____ │
│ _____ │ │ _____   │ │   _____ │
│ _____ │ │ _____   │ │ 2. _____ │
│ _____ │ │ _____   │ │   _____ │
│ _____ │ │   │ │   _____ │
│   │ │   │ │ 3. _____ │
│   │ │   │ │   _____ │
│   │ │   │ │   _____ │
└────────────────────┘ └────────────────────┘ └──────────────────────────┘

B. Detailed Understanding

I. Tell if the following statements are true (T) or false (F) according to the text.

1. _____ The giant Leshan statue was once thought to be the tallest Buddha statue in the world.

2. _____ The feet of the giant Leshan statue are threatened by the pounding water.

3. _____ The fund supporting the facelift project is provided by the Chinese government.

4. _____ UNESCO will investigate into the protection of the giant Leshan statue this year.

5. _____ Visitors have done damage to the giant Leshan statue for years.

6. _____ There are more than four world natural and cultural heritage sites in Beijing.

7. _____ Sichuan's regulation specially drafted for protecting local world heritage sites is the first law of its kind in the world.

8. _____ In Sichuan, developers are encouraged to build hotels and other related facilities in the core area of world heritage sites so as to promote tourism.

9. _____ The protection of ecological environment is helpful in the preservation of world heritage sites.

10. _____ How domestic waste should be treated is not stated in the regulation.

II. Explain the sentences by filling in the blanks.

1. **Text sentence:** Sichuan takes pains to protect world heritage sites.

   Interpretation: Sichuan _____ protecting world heritage sites.

2. **Text sentence:** The world's tallest Buddha statue... is undergoing its second round of facelift...

    Interpretation: This is the second time _____...

3. **Text sentence:** ... who heads a 20-strong team to do the work.

    Interpretation: ... who is in charge of a team _____...

4. **Text sentence:** ... The giant Leshan statue has suffered... wear and tear by visitors for years.

    Interpretation: ... visitors _____ for years.

5. **Text sentence:** China boasts a total of 28 world natural and cultural heritage sites.

    Interpretation: _____ 28 world natural and cultural heritage sites.

6. **Text sentence:** Four are in southwestern Sichuan Province, second only to the capital Beijing.

    Interpretation: Four are in southwestern Sichuan Province and only the capital Beijing _____.

7. **Text sentence:** ... the first local law of its kind enacted in China.

    Interpretation: ... the first _____ in China.

8. **Text sentence:** ... calling for enhanced efforts to protect ecological environment...

    Interpretation: ... asking people to _____ to protect ecological environment...

9. **Text sentence:** ... the wide use of environment-friendly vehicles with clean fuel...

    Interpretation: ... the wide use of vehicles which _____...

C. Detailed Study of the Text

1   **The project aims to make the statue, which is more than 1,200 years old, weatherproof, by installing drainage devices and protecting the statues feet against pounding waves.** (Para. 2) 这项工程是要给佛像安装排水设施，使其脚部能抵御汹涌波浪的冲击，从而使这座有着 1,200 多年历史的大佛能经受住风雨的侵蚀。

    乐山大佛位于我国四川省乐山市城东岷江、青衣江、大渡河三江汇合处，是倚凌云山栖霞峰临江峭壁凿造的一尊弥勒坐像。佛像的双脚浸在水中，句中 pounding waves 指的是崖壁下湍急的江水产生的对佛像脚部猛烈击打的波浪。

2　The Buddha statue, sitting squarely on a cliff, 71 meters from top to bottom and 28 meters across, is 18 meters higher than the standing Buddha statue in Bamian Valley, Afghanistan, once thought to be the tallest in the world. (Para. 6) 端坐在悬崖边的这座佛像高 71 米，宽 28 米。过去认为世界上最高的佛像是位于阿富汗巴米扬河谷的大立佛，而乐山大佛比它还要高出 18 米。

once thought to be the tallest in the world 在句中作定语，修饰 the standing Buddha statue in Bamian Valley, Afghanistan。应当理解为：The standing Buddha statue in Bamian Valley, Afghanistan, was once thought to be the tallest in the world.

句中提到的巴米扬大立佛由两座具有 1,500 年以上历史的佛像组成，是世界上最高的立式石雕佛像。它们于 2001 年 3 月被阿富汗塔利班军方以净化穆斯林环境为由炸毁，此举激起了国际社会的广泛斥责。

3　The province has put into effect a regulation specially drafted for protecting local world heritage sites, the first local law of its kind enacted in China and one of the few similar regulations in the world. (Para. 8) 该省已专门制定并实施了有关保护当地世界遗产的法规。制定此类地方法规在中国尚属首次，在世界范围内也较为罕见。

句中 the first local law of its kind enacted in China 和 one of the few similar regulations in the world 都是作前面 regulation specially drafted for protecting local world heritage sites 的同位语。

4　... the wide use of environment-friendly vehicles with clean fuel in and around world heritage sites. (Para. 9) 在世界遗产地及周边地区广泛使用消耗清洁燃料的环保车辆。

environment-friendly vehicles 指的是使用天然气、电力、氢燃料等清洁燃料的环保车辆。这类车辆废气排放量低，对环境污染较少，也就有利于对世界遗产地的保护。

5　The regulation also strictly specifies the discharge of polluted water, smoke and dust as well as the treatment of domestic waste. (Para. 9) 法规还严格限定了污水、烟和灰尘的排放标准以及对生活垃圾的处理要求。

句中 as well as 连接的并列成分是 the discharge of polluted water, smoke and dust 和 the treatment of domestic waste，它们都是 specifies 的宾语。

D. Talking About the Text

Work in pairs. Ask and answer the following questions first and then put your answers together to make an oral composition.

1. How many world heritage sites are there in Sichuan Province?
2. Which one is the world's tallest Buddha statue?
3. What is being done to protect this statue?
4. Where does the cost come from?
5. What did the Chinese government do to protect the statue last year?

6.   Besides repair and restoration work, what else has Sichuan Province done to protect the local world heritage sites?

7.   What does the regulation ban?

8.   What does the regulation call for?

9.   What else is also specified in the regulation?

10.   What is the result of all these efforts?

E. Vocabulary Practice

I.   **Fill in the blanks with the new words or expressions from Text A.**

1.   The insurance policy does not cover the damage caused by normal _____ and tear.

2.   If the number of wolves keeps increasing, the _____ balance of the area will be destroyed.

3.   _____ speaking, this book is not a novel but only a short story.

4.   Compared with oil and coal, natural gas is a kind of clean and cheap _____.

5.   As a student, you must obey the rules and _____ of the school.

6.   The museum has _____ a series of facelifts and will be re-opened in one week.

7.   The doctor will _____ when and how often you should take the medicine.

8.   There is a ban on the _____ of poisonous waste into the river.

9.   The Chairman was so busy that he asked his secretary to _____ the speech for him.

10.   This is a _____ designed diet for those who want to lose weight.

II.   **Complete the following dialogs with appropriate words expressions from Text A.**

1.   A: How did Henry do in the game?
     B: He lost in the first _____, but won in the second and third.

2.   A: Can you tell me something about _____?
     B: It's the art and study of designing buildings.

3.   A: Have you heard of the Temple of Heaven?
     B: Yes, of course. It was included on the World Cultural _____ list in 1998.

4.   A: It's said that the new traffic regulations have been _____.
     B: They will be put into effect from next month on.

5.   A: Do you really love me?
     B: Yes, I love you from the _____ of my heart.

6.   A: I can't understand why strong earthquakes hit Japan more than ten times during the past century.

B: One of the reasons is that the country is located in an earthquake _____.

7. A: The church was almost destroyed during World War II.

    B: But now it has been _____ to the way it looked in 1910 when it was first built.

8. A: Why did you say that the last moment of the football match was very exciting?

    B: Because the ball hit _____ on the post but then it bounced into the Net!

9. A: The hotel has excellent _____.

    B: But the service is far from satisfactory.

10. A: Every time it rains heavily, the water on the ground would be deep enough to drown a car!

    B: Oh, my God! Measures should be taken to improve the _____ system in the area.

Text B

 Before Reading

Discuss the following questions in class.

1. What do you know about the World Heritage List?

2. If a site in the city where you live has recently been included on the List, how would you feel about it?

 Reading

World Heritage Status—a Mixed Blessing

1   What do the Statue of Liberty and the Ming Tombs have in common? These unique wonders belong not to any one nation but to all humankind[1] as internationally protected sites of outstanding universal[2] value. That, at least, is the guiding principle behind the World Heritage Convention[3], a treaty[4] directed by the United Nations Educational, Scientific and Cultural Organization (UNESCO) for the protection of our common natural and cultural inheritance[5].

2   Making the list is not an easy task. A nation must actively promote a potential site for inclusion[6] under the Convention. Beyond demonstrating[7] a site's outstanding

universal value, a country must define the boundaries of the site and provide a detailed long-term management plan. Only a handful make the cut[8] each year.

3   What's the payoff[9]?

4   Conservationists[10] say the international backing[11] of the World Heritage Convention is a valuable aid in promoting conservation[12] initiatives[13], a bonus[14] for tourism, and a source of national pride. For example, one research finds that global recognition has encouraged Belizeans to take a more active role in protecting their natural heritage.

5   Although the World Heritage Convention can provide a powerful aid in protection, making the list can be a mixed blessing. With listed status comes international exposure[15]. Tourists eager[16] to see the wonders of a site are quickly followed—or in some cases preceded[17]—by developers and others anxious[18] to exploit[19] the money they spend. Countries like Belize face the challenge to avoid the types of problems a massive[20] influx[21] of tourism and recognition has brought to other World Heritage sites.

6   Although listing requires detailed tourism and site management plans, often no amount of planning can be enough. At Xi'an, China, site of the famous Terra-cotta Warriors, a poorly situated[22] new museum to handle the crowds may in fact have a negative impact on the site. On the Belizean Barrier Reef, developers are closing in, exploiting the World Heritage status a few miles away to sell land to prospective[23] customers over the Internet.

7   All nations that sign the Convention commit financial[24] resources to protect and promote their own sites and to help—when possible—threatened sites around the world. Of 730 sites in 125 countries, 33 are currently[25] included on the formal World Heritage List of Sites in Danger. The special status provides threatened[26] sites access to emergency[27] funding. Endangered sites face numerous threats, from natural disasters[28], pollution, and lack of funding to war. The threat of removal[29] from the list can also serve as a powerful incentive[30] for nations to be careful about protecting their sites.

New Words and Expressions

1   humankind /ˌhjuːmənˈkaɪnd/ *n.* （统称）人，人类
2   universal * /ˌjuːnɪˈvɜːsəl/ *a.* 普遍的，共同的
3   convention ▲ /kənˈvenʃən/ *n.* 公约，协议
4   treaty * /ˈtriːtɪ/ *n.* （尤指国家间的）条约，协定
5   inheritance # /ɪnˈherɪtəns/ *n.* 从自然界承袭的共有资产（指水、土地、空气等）

6　inclusion# /ɪn'kluːʒən/ n. （被）包括，包含

7　demonstrate ★ /'demənstreɪt/ v. 证明，说明

8　make the cut 达到标准

9　payoff# /'peɪɒf/ n. 回报，报偿

10　conservationist# /ˌkɒnsə'veɪʃənɪst/ n. 自然环境保护主义者

11　backing /'bækɪŋ/ n. 支持

12　conservation ★ /ˌkɒnsə'veɪʃən/ n. （对自然资源的）保护

13　initiative /ɪ'nɪʃɪətɪv/ n. 主动的行动

14　bonus ★ /'bəʊnəs/ n. 额外给予的东西

15　exposure ★ /ɪk'spəʊʒə(r)/ n. 暴露

16　eager /'iːɡə(r)/ a. 热切的，渴望的

17　precede■ /ˌpriː'siːd/ vt. 在……之前发生（或出现）

18　anxious /'æŋkʃəs/ a. 渴望的，急切的

19　exploit ★ /ɪk'splɔɪt/ vt. （为获取利益而）利用

20　massive ▲ /'mæsɪv/ a. 大量的，大规模的

21　influx■ /'ɪnflʌks/ n. （人、资金或事物的）涌入，流入

22　situated# /'sɪtjueɪtɪd/ a. （尤指财政方面的）处于……境地的

23　prospective /prəʊ'spektɪv/ a. （仅用于名词前）可能的，有希望的

24　financial ★ /faɪ'nænʃəl/ a. 财政的，金融的

25　currently# /'kʌrəntlɪ/ ad. 现在，目前

26　threaten ★ /'θretən/ vt. 危及，对……构成威胁

27　emergency /ɪ'mɜːdʒənsɪ/ a. （仅用于名词前）应急的

28　disaster ★ /dɪ'zɑːstə(r)/ n. 灾难，大祸

29　removal ▲ /rɪ'muːvəl/ n. 除去，消除

30　incentive■ /ɪn'sentɪv/ n. 刺激，鼓励

Proper Nouns

Statue of Liberty 自由女神像 [美国纽约雕塑名]

the Ming Tombs 中国明代皇陵

Belize /be'liːz/ 伯利兹 [拉丁美洲国家]

Belizean /be'liːzɪən/ 伯利兹的，伯利兹人

Belizean Barrier Reef 伯利兹堡礁

Terra-cotta Warriors 兵马俑

After Reading

A. Main Idea

- -

Complete the following diagram with the expressions given below.

1. promotion of conservation initiatives
2. defining the boundaries of the site
3. problems resulted from a massive influx of tourists and recognition
4. a source of national pride
5. demonstrating a site's "outstanding universal value"
6. negative impacts on the sites due to inadequate planning of site management
7. a bonus for tourism
8. providing a detailed long-term management plan
9. access to financial resources

It's not easy to be included on the World Heritage List. A country must make the following preparations:

1. _____
2. _____
3. _____

↓

Making the list can be a mixed blessing.

Listing brings benefits:

1. _____
2. _____
3. _____
4. _____

World Heritage status can pose problems as well:

1. _____
   _____
2. _____
   _____

B. Detailed Understanding

I. Make correct statements according to the text by combining appropriate sentence parts in Column A with those in Column B.

Column A	Column B
1. What all world heritage sites have in common is that _____.	a. define the boundaries of the site and provide a detailed long-term management plan
2. The World Heritage Convention is a treaty _____.	b. directed by the UNESCO for the protection of our common natural and cultural inheritance
3. Although the international support of the World Heritage Convention is very helpful, _____.	c. a massive influx of visitors and recognition may bring to its world heritage sites
4. To apply for world heritage status, a country must _____.	d. it may be removed from the World Heritage List
5. It's a challenge for a country to avoid the problems _____.	e. they have "outstanding universal values"
6. Because the Belizean Barrier Reef has been included on the list, Belizeans _____.	f. feel proud of their nation and make more efforts to protect their natural heritage
7. Since the new museum is poorly situated to handle the crowds _____.	g. emergency funding and extra help
8. In Belize, developers try to make money by _____.	h. the heritage status can be a mixed blessing
9. UNESCO provides threatened sites with _____.	i. it may do harm to the site of Terra-cotta Warriors
10. If a nation is not careful enough about protecting its world heritage sites, _____.	j. selling the land a few miles away from the Belizean Barrier Reef over the Internet

II. Explain the sentences by filling in the blanks.

1. **Text sentence:** What do the Statue of Liberty and the Ming Tombs have in common?

   Interpretation: What are the _____?

2. **Text sentence:** Making the list is not an easy task.

   Interpretation: It's not easy to _____.

3. **Text sentence:** Beyond demonstrating a site's outstanding universal value...

   Interpretation: _____ a site is of outstanding universal value...

4. **Text sentence:** Only a handful make the cut each year.

   Interpretation: Only a few of the sites _____.

5. **Text sentence:** ... making the list can be a mixed blessing.

   Interpretation: ... there are _____ in getting the world heritage status.

6. **Text sentence:** With listed status comes international exposure.

   Interpretation: If a site is included on the World Heritage List, it is then _____.

7. **Text sentence:** ... often no amount of planning can be enough.

   Interpretation: ... often no matter _____, it's still not enough.

8. **Text sentence:** ... commit financial resources to protect and promote their own sites...

   Interpretation: ... provide _____...

9. **Text sentence:** The special status provides threatened sites access to emergency funding.

   Interpretation: With this special status, threatened sites _____.

10. **Text sentence:** The threat of removal from the list can serve as a powerful incentive for nations to be careful about protecting their sites.

    Interpretation: Afraid of being removed from the list, nations will _____.

C. Detailed Study of the Text

----

1 **Only a handful make the cut each year.** (Para. 2) 每年只有少数申请地能入选世界遗产名单。
句子补充完整了是：Only a handful of the sites make the cut each year.

2　**With listed status comes international exposure.** (Para. 5) 榜上有名后，这一地区便为世人所知了。

这是一个倒装句，正常语序为：International exposure comes with listed status.

3　**Countries like Belize face the challenge to avoid the types of problems a massive influx of tourism and recognition has brought to other World Heritage sites.** (Para. 5) 被列入世界遗产名单后所获得的广泛认可和大量旅游者的涌入已经给其他世界遗产地带来各类问题，如何避免这些问题是诸如伯利兹等国所面临的挑战。

句中 a massive influx of tourism and recognition has brought to other World Heritage sites 是定语从句，修饰前面的 problems。从句的主语是 a massive influx of tourism and recognition。

4　**On the Belizean Barrier Reef, developers are closing in, exploiting World Heritage status a few miles away to sell land to prospective customers over the Internet.** (Para. 6) 在伯利兹堡礁，开发商们借着世界遗产地的名分，在几英里远的地方搞开发，在互联网上向可能的客商兜售土地。

伯利兹堡礁位于中美洲的伯利兹海域，是西半球最大的珊瑚群，于 1996 年被列入世界自然遗产地名单。

5　**All nations that sign the convention commit financial resources to protect and promote their own sites and to help—when possible—threatened sites around the world.** (Para. 7) 所有签署《世界遗产公约》的国家都承诺提供资金对本国世界遗产地进行保护和推广宣传。如果情况许可的话，也对世界其他濒临毁坏的遗产地提供帮助。

句中 when possible 补充完整了是：when it is possible to do so = when it is possible to help threatened sites around the world.

D. Further Work on the Text

- -

Write down at least three more comprehension questions of your own. Work in pairs and ask each other these questions. If you can't answer any of these questions, ask your classmates or the teacher for help.

1. _____

2. _____

3. _____

E. Vocabulary Practice

- -

I. **Find the word that does NOT belong to each group.**

1. A. treaty   B. crowd   C. tomb   D. boundary
2. A. initiative   B. negative   C. prospective   D. active
3. A. humankind   B. payoff   C. inheritance   D. long-term
4. A. global   B. national   C. universal   D. international
5. A. pollution   B. convention   C. protection   D. recognition

II. **Complete the following sentences with appropriate words in their correct form.**

1. **liberty, liberal, liberate**
   1) China needs more _____ trade relations with Europe.
   2) They fought for justice and _____.
   3) After the 8-day battle, the village was finally _____ from the enemy occupation.

2. **tour (*n.*), tour (*v.*), tourist, tourism**
   1) The World Heritage status will lead to increased _____ in the area.
   2) She is a very good _____ guide.
   3) The Summer Palace is a popular _____ attraction.
   4) They spent four weeks _____ Europe.

3. **hand (*n.*), hand (*v.*), handful**
   1) _____ me the dictionary, please.
   2) We are short of _____ now.
   3) Only a _____ of people came to attend the meeting.
   4) Would you mind giving me a _____ with these bags?
   5) The hour _____ of the clock needs repairing.

4. **danger, dangerous, endanger, endangered**
   1) The sea turtle is a(n) _____ species.
   2) The current situation is highly _____.
   3) Doctors said he is now out of _____.
   4) You will _____ your health if you work too hard.

5. **removal, remove, remover, removable**
   1) Soldiers carefully _____ the little girl from the collapsed building.
   2) You'd better use stain _____. Otherwise the oil stains on the clothes won't come off easily.
   3) Tom is very interested in the toy car because every part of it is _____.
   4) The president has called for the _____ of all foreign troops in the country.

Word Study

condition

n. 1. 状况；状态：This used car is in good condition. 这辆旧车的车况良好。/She is overweight and out of condition. 她体重超重，健康状况不佳。

2. [*pl.*] 环境：The plants grow best in warm, dry conditions. 这种植物最适合在温暖、干燥的环境下生长。

3. 条件：They have agreed to stop striking if their conditions are met. 他们已经同意只要提出的条件得到满足就停止罢工。

on condition that 如果：You can go out on condition that you must come back home before 11 o'clock. 你可以出去，条件是 11 点之前必须回家。

派 conditional

detail

n. 1. 细节；详情：For further details, please write to our company. 欲知详细情况，请致函本公司。

2. 枝节；琐事：She doesn't like the details of housekeeping. 她不喜欢琐碎的家务。

vt. 详述；详细说明：In a special report the newspaper detailed the traffic accident. 那份报纸在一篇特约报道中详述了这起交通事故的情况。

in detail 详细地：The issue will be discussed in detail at the next meeting. 这个问题将在下次会议上详细讨论。

派 detailed

suffer

vt. 经受；遭受；蒙受：The party suffered a defeat in the general election. 该党在大选中失败。

vi. 1. (*from*) 受痛苦；受疼痛；受苦难；患病：The old man suffered from the loss of memory. 这位老人患了健忘症。

2. 受损害；受损失：Business has suffered a lot since the earthquake. 地震以后，营业受到不小的损失。

派 sufferer, suffering

handle

n. 柄；把手：He turned the handle and opened the door. 他转动把手，打开了门。

vt. 1. 处理；应付；对待：I don't know how to handle such problems in the right way. 我不知道如何正确处理此类问题。

2. 操作：Workers are handling the machines with ease and skills. 工人们正熟练地操作着机器。

派　handling

negative

a. 1. 否定的：They gave a negative answer to my request. 他们对我的请求给予了否定的答复。

2. 反面的；消极的：Some researchers have a negative attitude to the experiment. 一些研究人员对这项实验持消极态度。

3. 负的；阴性的：Students are learning what a negative current is. 学生们正在学习什么是负电流。

n. 1. 底片：Don't forget to have these negatives developed this afternoon. 别忘了今天下午去冲洗这些底片。

2. 负数：Two negatives make a positive. 负负得正。

派　negatively

Writing Practice

I. Edit the punctuations in the following sentences.

1. The accessibility of the computer, has increased tremendously over the past several years.

2. What has humanity done about the growing concern of global warming.

3. People continue to worry about the future, our failure to conserve resources has put the world at risk.

4. The professor has given me three options; to retake the exam, to accept the extra credit assignment, or to fail the course.

5. The Easter basket contained—Easter eggs, chocolate rabbits, and candies.

6. He lived to be one hundred twenty one.

7. An introductory clause is a brief phrase that comes, yes, you guessed it at the beginning of a sentence.

8. These kids test scores are the highest in the nation.

9. Bill Gates CEO of Microsoft is the developer of the operating system known as Windows.

10. The computer store was filled with video games computer hardware and other electronic devices.

11. Los Angeles, CA is one of the largest cities in the United States.

12. Ryan went to the beach yesterday but he forgot his sunglasses.

13. While I was at his house, John asked: Do you want anything to eat?

14. Lily could you come here for a moment?

15. To register, you will need your drivers license and-or your birth ID.

II. Punctuate the following paragraph.

There I was in the summer of 1986 a member of the Santa Clara Fire Department in search of a building to burn down for training purposes I found an abandoned city building that fit the bill It used to be part of a hospital and was equipped with an X-ray machine surgical supplies and oxygen bottles all of which had to be removed before we set fire to the building The Director of Emergency Services said the equipment in the building was too outdated to be of use so I recommended they be donated to an organization I had heard good comments about: Refugee Relief International Fine said the Director You make the arrangements

III. Rewrite the sentences according to the models.

Model A:

Original sentence: Four are in southwestern Sichuan Province and only the capital Beijing has more world heritage sites.

New sentence: Four are in southwestern Sichuan Province, second only to the capital Beijing. (Text A)

1. Tom was a fast runner and only Peter ran faster than he in the race.

_____

2. Jackie Chen is a popular film star and only Tom Hanks is more popular than him.

_____

Model B:

Original sentence: As this regulation has been put into effect, Sichuan Province is aiming to preserve its world heritage sites for future generations.

New sentence: With this regulation in effect, Sichuan Province is aiming to preserve its world heritage sites for future generations. (Text A)

3. As the wildlife has been put under protection, we hope the ecological environment will be improved.

_____

4. As children's lives are already in danger, we should try our best to save them.

_____

Model C:

Original sentence: If a site is included on the list, it's open to the whole world.

   New sentence: With listed status comes international exposure. (Text B)

5. If you win the first prize, you will get both fame and money.

   _____

6. Once you get the certificate, there will be chances to get promoted.

   _____

Model D:

Original sentence: All nations that sign the Convention commit financial resources to... help
                  threatened sites around the world if it's possible to do so.

   New sentence: All nations that sign the Convention commit financial resources... to help—
                  when possible—threatened sites around the world. (Text B)

7. I'll go to see you once a week when it's possible for me to do so.

   _____

8. You can ask for help when it's necessary to do so.

   _____

IV. Combine each set of the sentences into one, using the connective words or expressions provided.

1. a. The statue is more than 1,200 years old.

   b. The project aims to make the statue weatherproof.

   c. Drainage devices will be installed.

   d. The statue's feet will be protected against pounding waves.

   New sentence: _____ (which, by, and)

       (Text A)

2. a. According to Zeng Zhiliang, the facelift project is the first one in China to use a World Bank loan to repair and restore a world heritage site.

   b. Zeng Zhiliang is an engineer in the field of ancient relics and architecture.

   c. Zeng Zhiliang heads a 20-strong team to do the work.

   New sentence: _____ (who) (Text A)

3. a. The giant Leshan statue has suffered apparent damage from the wind, water, acid rains, and wear and tear by visitors for years.

   b. The giant Leshan statue was included on the World Cultural Heritage List in 1996.

   New sentence: _____ (which) (Text A)

4. a. The Buddha statue sits squarely on a cliff.

   b. It is 71 meters from top to bottom and 28 meters across.

c. It is 18 meters higher than the standing Buddha statue in Bamian Valley, Afghanistan.

d. The standing Buddha statue in Bamian Valley was once thought to be the tallest in the world.

New sentence: _____ (-ing) (Text A)

5. a. The World Heritage Convention can provide a powerful aid in protection.

b. Making the list can be a mixed blessing.

New sentence: _____ (although) (Text B)

6. a. Listing requires detailed tourism and site management plans.

b. Often no amount of planning can be enough.

New sentence: _____ (although) (Text B)

7. a. On the Belizean Barrier Reef, developers are closing in.

b. They are exploiting the World Heritage status a few miles away.

c. They want to sell land to prospective customers over the Internet.

New sentence: _____ (-ing) (Text B)

8. a. Some nations have signed the convention.

b. All of these nations commit financial resources to protect and promote their own sites and to help threatened sites around the world.

c. When possible, they will also help threatened sites around the world.

New sentence: _____ (that) (Text A)

V. Translate the following sentences into English.

1. 他们费尽心思对计划保守秘密。（take pains）
2. 我们的产品必须经久耐用。（wear and tear）
3. 由纽约市市长率领的代表团将于明日访问我市。（head）
4. 去年她母亲动了一系列手术。（undergo）
5. 每年有数百名艺术家竞争这些奖项，最终只有10名获奖。（make the cut）
6. 他利用父亲的名声为自己找了份好工作。（exploit）
7. 计算机的发明对人们的生活产生了巨大的影响。（impact）
8. 警察认为这种惩罚能起到警示他人的作用。（serve）

VI. Translate the following paragraph into Chinese.

World Heritage is the superb crystallization (结晶) of every country's history, culture and civilization. Both natural heritage and cultural heritage constitute precious common property of mankind. We have noted that numerous world heritage sites have been well protected, thanks to the publicity and education on the part of the local governments. Meanwhile, however, we have also noted with regret that some of the heritage sites have been irreversibly (不可逆转地)

damaged due to wars, tourism development, natural disasters, environmental pollution, shortage of funds and people's lack of knowledge, which might deprive us of the diversified species, the beautiful landscapes and the splendid cultures created by our ancestors.

VII. Practical English Writing

Directions: Now you are a member of a committee that has been set up by the government for the protection of a historical, cultural, or scenic site in your hometown. You are responsible for writing a plan to attract funding for your project. In your plan, you will give a brief description of the site you want to protect and then explain how you are going to use the fund for the project. Before you write your plan, you may discuss with your classmates and jot down a few notes in the following table.

Description of the Site: (Its location and surroundings; its physical features; its history or major historical figures and events associated with it; cultural figures or works associated with it; etc.)	
Uses of the Fund: (Restoration; maintenance; moving residents out of the site; promotion; etc.)	

You can begin your plan this way:

A Plan for the Protection of _____

In the past two decades, we have clearly achieved rapid progress in the economic development of our city. However, relatively few efforts have been made for the protection of the splendid historical/cultural/scenic sites of the city. For example,

_____

_____

_____

_____

_____

Text A

 Before Reading

Discuss the following questions in class.

1. What harm do wars bring to the world in general?
2. How much do you know about the war against Iraq?

 Reading

Open Letter and Appeal[1]

To: President John Sweeney and the General Executive Council of the National Workers Association (NWA)

1 Across the country, local, district[2], and national unions, labor councils, state labor federations[3] and numerous other labor organizations representing millions of working people have adopted resolutions[4] condemning[5] the war in Iraq, and calling for an end to the occupation[6] and return of all troops[7] to their homes and families.

2 110 of these organizations have banded[8] together to form U.S. Labor Against the War (USLAW), a national organization committed[9] to ending the war, returning the troops, restoring funding to social programs and government services, and changing the direction of U.S. foreign policy.

3 Union members and their family members are being killed, wounded, and disabled in a war that has already killed almost 1,500 U.S. military personnel, wounded more than 10,500 others, a war in which more than 100,000 Iraqi civilians[10] have died. This war is draining[11] away resources from our communities, starving[12] or eliminating[13] essential public services and social programs, endangering our democratic[14] rights, and making our country even less secure[15].

4 It is time for labor to speak out! At this time of discussing about renewing[16] our labor movement, how can we not discuss the most urgent[17] issue facing America and its working families? We ask you to put the issue of the war on the agenda of the upcoming Executive Council meeting. And we urge[18] the national leadership of the NWA to oppose this wild, illegal and immoral[19] war.

5 More specifically[20], we ask for action on the following proposals[21] by the Executive Council and the annual convention of the NWA.

6   The NWA should demand an immediate end to the U.S. occupation of Iraq and return of U.S. troops to their homes and families, and the reordering[22] of national priorities[23] toward peace and meeting the human needs of our people.

7   Through its community service programs, the NWA and its state and local organizations should assist union members and their families who are called upon to serve in the armed forces and returning veterans by identifying and providing information about resources and services available to meet their needs, by advocating[24] for their interests, and by protecting their jobs and benefits and those of unorganized workers in similar circumstances[25].

8   Sisters and brothers, this war is draining away precious resources essential to meet the human needs of working and poor people. It is isolating[26] the U.S. from the community of nations and provoking[27] the spread of terrorism. It is weakening rather than reinforcing[28] the rule of international law. It has led to an erosion[29] of our most basic rights and liberties. And it is doing terrible and direct harm to many thousands of military families.

9   We, the American labor movement, should take a stand and speak out on the biggest issue facing working people and the country as a whole. We urge you to join us!

10   Labor leaders and organization signers[30] are invited to send their names, titles, and organizations (whether signing for an organization, or as an individual with an organization for identification[31] only) to USLAW.

New Words and Expressions

1   apppeal /ə'pi:l/ n. 呼吁，恳请
2   district ★ /'dɪstrɪkt/ n. 区，地区，行政区
3   federation ■ /ˌfedə'reɪʃn/ n. 联合会
4   resolution ★ /ˌrezə'lu:ʃən/ n. 正式决定，决议
5   condemn ▲ /kən'dem/ vt. 谴责
6   occupation # /ˌɒkjʊ'peɪʃən/ n. 占领，占据
7   troop ★ /tru:p/ n. [常 pl.] 军队，部队
8   band ▲ /bænd/ v. 用带绑扎，结合起来
9   commit to 对……作出承诺，承担义务
10 civilian ▲ /sɪ'vɪljən/ n. 平民，百姓
11 drain ▲ /dreɪn/ vt. 使渐渐耗尽
     drain away （使）用尽
12 starve ★ /stɑːv/ v. 使极度缺乏；（使）挨饿
13 eliminate ★ /ɪ'lɪmɪneɪt/ vt. 根除，消除，排除

14　democratic ★ /ˌdeməˈkrætɪk/ *a.* 民主的，有民主精神（或作风）的

15　secure ▲ /sɪˈkjʊə(r)/ *a.* 安全的，无危险的

16　renew ▲ /rɪˈnjuː/ *v.* 重新开始，继续

17　urgent /ˈɜːdʒənt/ *a.* 紧迫的，紧急的

18　urge /ɜːdʒ/ *vt.* 竭力主张，强烈要求

19　immoral # /ɪˈmɒrəl/ *a.* 不道德的，邪恶的

20　specifically ▲ /spɪˈsɪfɪkəlɪ/ *ad.* 特别地；明确地

21　proposal /prəʊˈpəʊzəl/ *n.* 提议，建议

22　reorder # /riːˈɔːdə(r)/ *v.* 重新安排，重新整理

23　priority /praɪˈɒrətɪ/ *n.* 优先权

24　advocate ▲ /ˈædvəkeɪt/ *vt.* 拥护，提倡，主张

25　circumstance ★ /ˈsɜːkəmstəns/ *n.* [*pl.*] 境遇，境况，经济状况

26　isolate /ˈaɪsəleɪt/ *v.* 使孤立，使隔离

27　provoke ★ /prəʊˈvəʊk/ *vt.* 激起，引起

28　reinforce /ˌriːɪnˈfɔːs/ *v.* 强化

29　erosion ▲ /ɪˈrəʊʒən/ *n.* 削弱，减少

30　signer # /ˈsaɪnə(r)/ *n.* 签名人，签字者

31　identification # /aɪˌdentɪfɪˈkeɪʃən/ *n.* 身份证明

Proper Nouns

Iraq /ɪˈrɑːk/ 伊拉克 [西南亚国家]

Iraqi /ɪˈrɑːkɪ/ *a.* 伊拉克的

John Sweeney /ˈswiːnɪ/ 约翰·斯威尼，美国最大工会组织美国劳工联盟及产业工会联合会（劳联-产联）主席

 ## After Reading

A. Main Idea

- -

Complete the following diagram with the sentences or expressions given below.

1.　It is doing terrible and direct harm to many thousands of military families. It has already killed almost 1,500 U.S. military personnel, wounded more than 10,500 others.

2.　The NWA should demand an immediate end to the U.S. occupation of Iraq and return of U.S. troops to their homes and families.

3. The NWA and its state and local organizations should assist union members and their families who are called upon to serve in the armed forces and returning veterans.

4. More than 100,000 Iraqi civilians have died in the war.

5. calling for an end to the occupation and return of all troops to their homes and families

6. thus weakening rather than reinforcing the rule of international law

7. 110 of these organizations have banded together to form U.S. Labor Against the War (USLAW).

8. The NWA should demand the reordering of national priorities toward peace and meeting the human needs of our people.

9. It is draining away resources from our communities, endangering people's most basic rights and liberties, and making the U.S. even less secure.

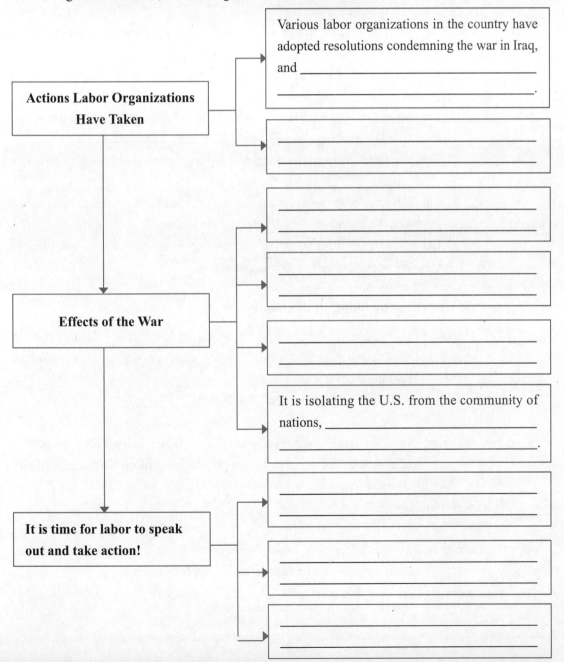

Actions Labor Organizations Have Taken

Various labor organizations in the country have adopted resolutions condemning the war in Iraq, and _____

_____.

Effects of the War

It is isolating the U.S. from the community of nations, _____

_____.

It is time for labor to speak out and take action!

B. Detailed Understanding

I. Tell if the following statements are true (T) or false (F) according to the text.

1. _____ Only local and district labor unions have united to adopt resolutions against the Iraqi war.

2. _____ USLAW has not yet been formally established so far but its mission is already made clear to the public.

3. _____ Great damage has been done to both countries involved in the war but comparatively speaking the Iraqis have suffered much more.

4. _____ The issue of war has been the most common one for American society and therefore deserves great concern.

5. _____ Enough proposals regarding the war have been put forward at the upcoming Executive Council meeting.

6. _____ The national leadership of the NWA should take action to oppose this unjust war.

7. _____ Union members and their family members have served in the armed forces and have killed or disabled almost 1,500 U.S. military personnel.

8. _____ This letter appeals to all members of the labor organizations to make their due contributions to the ending of the war.

9. _____ The U.S. has been separated from the international community as it has provoked the spread of terrorism throughout the world.

10. _____ Labor leaders and organization signers may send their names and titles to USLAW only on behalf of their organizations.

II. Explain the sentences by filling in the blanks.

1. **Text sentence:** One hundred and ten of these organizations have banded together to form U.S. Labor Against the War (USLAW), a national organization committed to ending the war...

   Interpretation: Among these organizations, 110 of them _____...

2. **Text sentence:** Union members and their family members are being killed, wounded, and disabled in a war that has already killed almost 1,500 U.S. military personnel...

   Interpretation: The Iraqi war, which has _____...

3. **Text sentence:** This war is draining away resources from our communities, starving or eliminating essential public services and social programs...

   Interpretation: This war has taken away so many resources from our communities that _____...

4. **Text sentence:** At this time of discussion about renewing our labor movement, how can we not discuss the most urgent issue facing America and its working families?

   Interpretation: While we _____.

5. **Text sentence:** We ask you to put the issue of the war on the agenda of the upcoming Executive Council meeting.

   Interpretation: We demand that _____.

6. **Text sentence:** More specifically, we ask for action on the following proposals by the Executive Council and the annual convention of the NWA.

   Interpretation: We particularly hope that _____.

7. **Text sentence:** The NWA should demand an immediate end to the U.S. occupation of Iraq and return of U.S. troops to their homes and families...

   Interpretation: The NWA should ask the U.S. government to _____...

8. **Text sentence:** It is isolating the U.S. from the community of nations and provoking the spread of terrorism.

   Interpretation: The war is separating the U.S. _____.

9. **Text sentence:** We, the American labor movement, should take a stand and speak out on the biggest issue facing working people and the country as a whole.

   Interpretation: As the American labor movement, we should _____ facing both the working people and the country as a whole.

C. Detailed Study of the Text

- -

1 **Across the country, local, district, and national unions, labor councils, state labor federations and numerous other labor organizations representing millions of working people have adopted resolutions condemning the war in Iraq, and calling for an end to the occupation and return of all troops to their homes and families.** (Para. 1) 全国各地代表数百万劳工的地方、地区及全国工会，劳工委员会，州劳工同盟以及其他众多劳工组织都已作出决议，谴责伊拉克战争，并呼吁停止占领，撤回所有部队，让他们回到家人的身边。这句话里有四个并列的主语，local, district, and national unions, labor councils, state labor federations 和 other labor organizations 后面的 representing... 与 condemning... 分别作为定语修饰主语和宾语部分，谓语动词为 adopted 。

美国劳工组织包括工会和工人政党两种组织形式，工联主义在其中有着强烈的影响。自1866 年起，先后有过全国劳工同盟、劳动骑士团、美国劳工联合会（简称"劳联"）、劳联–产联、劳工行动同盟等各种组织形式，其力量壮大，颇具影响。

Iraq 伊拉克，亚洲西南部国家。它是繁荣的古代美索不达米亚文明的所在地，先是落入阿拉伯人之手（7 世纪），而后又被奥斯曼土耳其人所占领（16 世纪）。1921 年它成为一个独立的王国，在费萨尔二世被暗杀后（1958 年）成为一个共和国。巴格达是其首都和最大的城市，人口约 15 万。

2　**... a national organization committed to ending the war, returning the troops, restoring funding to social programs and government services, and changing the direction of U.S. foreign policy. (Para. 2)** ……这一全国性组织致力于结束战争、撤回部队、恢复资助社会项目以及政府服务，改变美国外交政策的走向。

美国劳工反战联盟（USLAW）于 2003 年 1 月 11 日成立于芝加哥，其宗旨为结束战争、恢复和平，并帮助战后国家重建。

3　**... in a war that has already killed almost 1,500 U.S. military personnel, wounded more than 10,500 others, a war in which more than 100,000 Iraqi civilians have died. (Para. 3)** 在这场已经造成近 1,500 名美军官兵丧生，10,500 名士兵负伤，10 多万伊拉克平民丧生的战争中……

伊拉克战争从 2003 年 3 月 20 日开始，给美伊两国带来很大损失。文中所列伤亡人数截止至本文发表日期，即 2005 年 2 月 23 日。

4　**This war is draining away resources from our communities, starving or eliminating essential public services and social programs, endangering our democratic rights, and making our country even less secure. (Para. 3)** 这场战争正消耗着我们的社会资源，使重要的公共设施及社会项目资源匮乏，甚至无以为计，危害着我们的民主权利，并使国家更不安全。

这句话中有四个并列谓语：draining，starving or eliminating，endangering，making，用以加强语势，突出刻画战争的罪恶。

5　**... and the reordering of national priorities toward peace and meeting the human needs of our people. (Para. 8)** ……将国家优先发展重点重新转移到和平与满足人民基本需要上来。

美国对伊拉克的军事打击是自冷战结束以来美国发动的第 5 次对外重大军事行动，前 4 次对外重大军事行动为 1991 年的海湾战争、1998 年的"沙漠之狐"军事行动、1999 年的科索沃战争及 2001 年 10 月的阿富汗战争。这些行动引起美国国内及国际和平人士强烈不满，并给当地人民生活和美国经济发展带来负面影响，因而此信中将国家发展重点重新转移到和平和发展上的要求是由来已久的。

D. Talking About the Text

Work in pairs. Ask and answer the following questions first and then put your answers together to make an oral composition.

1. What have the unions, labor councils, federations and other organizations done in regard to the war in Iraq?
2. What have they called for?
3. Who established the organization USLAW and what is its objective?
4. What damage has the war caused to both countries?
5. What action should the leadership of the NWA take?
6. What else should the NWA and its state and local organizations do?
7. Why is all this needed?
8. In what other ways has the war done harm to thousands of military families and working people?
9. What does the war mean to the country?

E. Vocabulary Practice

I. **Fill in the blanks with the new words or expressions from Text A.**

1. He was forced to resign because of his _____ conduct.
2. The boy _____ the dog's neck with a yellow string.
3. Most people are willing to _____ violence of any sort.
4. They made a _____ to lose all the weight gained during the Christmas period.
5. Comprehensive control of soil _____ has brought noticeable achievements.
6. He has to move around in a wheelchair ever since the accident _____ him for work.
7. They were reluctant to _____ themselves to any definite proposals.
8. Now that prices are rising so fast, all my money is _____.
9. The air force has spared some airports for both military and _____ use to help the state develop civil aviation.
10. Such a questionable assertion is sure to _____ criticism.

II. **Complete the following dialogs with appropriate words or expressions from Text A.**

1. A: How was the competition?
    B: Awful. Our team was _____ in the first round.

2. A: If you have any objections, just _____.

   B: No. I'm all for it.

3. A: All such problems should be placed on our _____.

   B: You're right. We can't be too careful.

4. A: You can rely on me to _____ you any time you need.

   B: Thank you so much. You don't know what that means to me.

5. A: Kids today are so hard to control!

   B: I'll not _____ you.

6. A: Do you _____ banning cars in the city centre?

   B: Well, actually, I don't think I can make it without my car.

7. A: I guess she was just telling a lie by claiming to be ill.

   B: Never _____ opinions with facts.

8. A: I shall not agree with you under any _____.

   B: Just suit yourself.

9. A: He had come half a dozen times to _____ his sister.

   B: George's always been a thoughtful guy.

10. A: Mike, how many times do I have to tell you? You are at _____ to use MY car, not HIS!

    B: Sorry. I thought it was yours.

Text B

 Before Reading

Discuss the following questions in class.

1. What do you think is the most urgent issue facing a country immediately after a war?

2. Do you think it justifiable for a country to declare war on another and help with its reconstruction afterwards? Why or why not?

Reading

○　　　　○　　　　　　　　○　　　　○

Bush Wants to Shift $3.3 Billion to Strengthen[1] Iraq Security

1　　Washington—The Bush administration is preparing to seek approval[2] of Congress to shift $3.3 billion originally assigned to rebuild Iraq's shattered[3] infrastructure[4] into programs focused mainly on establishing law and order.

2　　The move comes against a background of steadily[5] worsening public security in the country as it approaches a crucial[6] first round of elections set for January.

3　　Those working on the changes said the proposed[7] reallocation[8] amounts to nearly one-fifth of the $18.4 billion Congress approved last November to rebuild Iraq. They said the shift would delay vital electricity[9], water and sewage[10] projects—all crucial to restoring Iraq's economy and building public support for the country's struggling interim[11] government.

4　　Instead, the money would go to a number of other programs, including $1.8 billion to strengthen the government's shaky[12] security organizations and additional funds to absorb unemployment. In a country where there are too many idle[13] men with little hope for work, job creation[14] is closely linked to improved security.

5　　At another level, however, the move indicates the administration's assessment[15] that substantial[16] changes are necessary to improve the security situations. Officials at the State Department working on the reconstruction[17] revisions said the shift in focus is part of a realization that funding even the most important projects makes little sense if conditions on the ground prevent their completion[18].

6　　"The first priority for our effort right now has to be security," Secretary[19] of State Colin Powell told reporters last week. "If a place is not safe to build a sewer[20] system, you can't spend the money."

7　　The revised[21] spending plan calls for adding 45,000 new recruits[22] to the national police force, raising 20 new battalions[23] to the current 42-battalion Iraqi National Guard[24], and beefing up[25] the force that protects Iraq's long borders in an attempt to stop the flow of illegal traffic.

8　　The shift of priorities, initially drawn up at the U.S. Embassy[26] in Baghdad, is expected to be reviewed at a meeting this week in Washington of senior officials from several government agencies[27] before going to Capitol Hill for final approval. There is no sign of major opposition[28] to the changes and those familiar with the issue predicted that Congress probably will approve the revised plan by the end of this month.

9   "There will be a lot of questions asked and there may be a hearing²⁹, but in general people want to be receptive³⁰ to this," said a congressional³¹ aide³². "There's a recognition that the realities (in Iraq) are much different than they were last fall" when Congress originally approved the budget³³.

New Words and Expressions

1   strengthen* /ˈstreŋθən/ v. 加强，巩固

2   approve /əˈpruːv/ v. 赞成，同意
    approval* /əˈpruːvəl/ n. 批准，认可

3   shatter■ /ˈʃætə(r)/ vt. 毁坏，使破灭

4   infrastructure /ˈɪnfrəˌstrʌktʃə(r)/ n. 基础设施

5   steadily /ˈstedɪlɪ/ ad. 持续地

6   crucial* /ˈkruːʃəl/ a. 至关重要的，决定性的

7   propose* /prəʊˈpəʊz/ v. 提议，建议，提出

8   reallocation■ /ˌriːæləʊˈkeɪʃən/ n. 重新分配

9   electricity /ˌɪlekˈtrɪsətɪ/ n. 电

10 sewage■ /ˈsjuːɪdʒ/ n. 下水道（系统）

11 interim■ /ˈɪntərɪm/ a. 临时的，暂时的

12 shaky# /ˈʃeɪkɪ/ a. 不可靠的，摇晃的，动摇的

13 idle* /ˈaɪdl/ a. 懒散的，无所事事的

14 creation /kriːˈeɪʃən/ n. 创造，创建

15 assessment# /əˈsesmənt/ n. 评估，评价

16 substantial* /səbˈstænʃəl/ a. 可观的，大量的

17 reconstruction# /ˌriːkənˈstrʌkʃən/ n. 重建，改造

18 completion# /kəmˈpliːʃən/ n. 实现，完成

19 secretary /ˈsekrətərɪ/ n. 部长，大臣

20 sewer■ /ˈsjʊə(r)/ n. 下水道

21 revise* /rɪˈvaɪz/ vt. 修订，修改

22 recruit▲ /rɪˈkruːt/ n. 新兵，新成员

23 battalion■ /bəˈtæljən/ n. 营；大队

24 guard /gɑːd/ n. 警卫员，看守，卫兵

25 beef up [口] 加强，补充（人数、兵力等）

26 embassy* /ˈembəsɪ/ n. 大使馆

27 agency* /ˈeɪdʒənsɪ/ n. （政府等的）专门行政部门

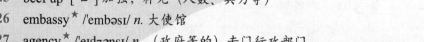

28 opposition# /ˌɒpəˈzɪʃən/ *n.* 反对，相反

29 hearing▲ /ˈhɪərɪŋ/ *n.* 意见（或证言）听取会

30 receptive# /rɪˈseptɪv/ *a.* （对建议等）愿接受的；易接受的

31 congressional# /kənˈgreʃənəl/ *a.* 国会的，大会的

32 aide■ /eɪd/ *n.* 助手，副官

33 budget /ˈbʌdʒɪt/ *n.* 预算

Proper Nouns

Secretary of State ［美国］国务卿

Baghdad /ˈbægdæd/ 巴格达 ［伊拉克首都］

Capitol Hill 美国国会山，（美国）国会

Colin Powell /ˈkɒlɪnˈpaʊəl/ 科林·鲍威尔 ［2001 年 1 月至 2005 年 1 月任美国国务卿，身跨军政两界，历经四届美国总统和无数次世界性危机］

 After Reading

A. Main Idea

Complete the following diagram with the sentences or expressions given below.

1. The overall attitude of the people towards the move is positive.

2. approve the revised plan by the end of this month

3. The shift is due to the worsening public security in Iraq.

4. the change in the administration's assessment and realization

5. at a meeting in Washington before going to Congress for final approval

6. The money would now go to other programs to improve security and reduce unemployment.

7. The Bush administration plans to shift $3.3 billion to strengthen Iraq's security.

8. The move involves recruiting new members to the police force and National Guard.

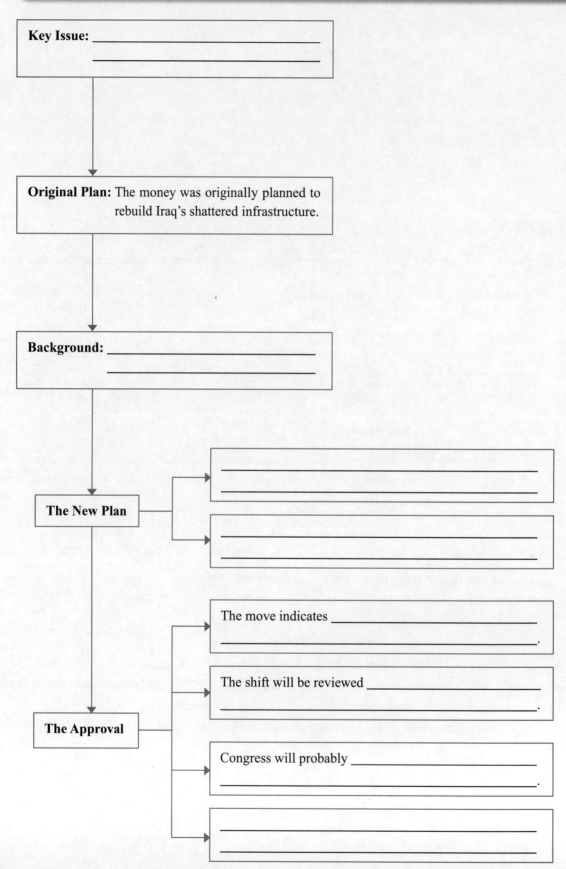

Key Issue: _____
_____

Original Plan: The money was originally planned to rebuild Iraq's shattered infrastructure.

Background: _____
_____

The New Plan
_____
_____

_____
_____

The Approval

The move indicates _____
_____.

The shift will be reviewed _____
_____.

Congress will probably _____
_____.

_____
_____

B. Detailed Understanding

I. Make correct statements according to the text by combining appropriate sentence parts in Column A with those in Column B.

Column A	Column B
1. Those working on the shift noted the changes would delay electricity, water and sewage projects _____.	a. from infrastructure rebuilding to security improvement
2. The security situation in Iraq is steadily worsening _____.	b. were approved by Congress last year
3. The total funds directed to Iraq that added up to $18.4 billion _____.	c. with the coming of the January election
4. Iraq's long borders would be more securely protected _____.	d. and thus affect economic restoration
5. The Bush administration as well as the public has already realized _____.	e. by regarding security as the first priority
6. Secretary of State Colin Powell shows that he is one of the firm supporters of the revised spending plan _____.	f. as employment is closely linked to improved security
7. The Bush administration intends to shift $3.3 billion _____.	g. and approval is most likely to be given by Congress
8. More job opportunities would be created with the shift _____.	h. the importance of improving the security situations in Iraq
9. So far there has been no major opposition from both the public and Congress _____.	i. if the plan of beefing up the police force could be approved by Congress

II. Explain the sentences by filling in the blanks.

1. **Text sentence:** The move comes against a background of steadily worsening public security in the country as it approaches a crucial first round of elections set for January.

   Interpretation: The proposal for the shift results from the fact that _____.

2. **Text sentence:** Instead, the money would go to a number of other programs, including $1.8 billion to strengthen the government's shaky security organizations and additional funds to absorb unemployment.

Interpretation: Instead, the money would be spent on other programs. For example, _____.

3. **Text sentence:** In a country where there are too many idle men with little hope for work, job creation is closely linked to improved security.

   Interpretation: In a country which is _____.

4. **Text sentence:** At another level, however, the move indicates the administration's assessment that substantial changes are necessary to improve the security situations.

   Interpretation: On the other hand, the shift shows _____.

5. **Text sentence:** The first priority for our effort right now has to be security.

   Interpretation: The most _____.

6. **Text sentence:** ... and beefing up the force that protects Iraq's long borders in an attempt to stop the flow of illegal traffic.

   Interpretation: ... and strengthening the armed forces _____.

7. **Text sentence:** The shift of priorities, initially drawn up at the U.S. Embassy in Baghdad, is expected to be reviewed at a meeting this week in Washington of senior officials from several government agencies before going to Capitol Hill for final approval.

   Interpretation: The change in the objectives of spending, which was first _____.

8. **Text sentence:** There is no sign of major opposition to the changes...

   Interpretation: There doesn't seem _____...

9. **Text sentence:** ... but in general people want to be receptive to this...

   Interpretation: ... but on the whole people _____...

C. Detailed Study of the Text

1 **The Bush administration is preparing to seek approval of Congress to shift $3.3 billion originally assigned to rebuild Iraq's shattered infrastructure into programs focused mainly on establishing law and order.** (Para.1) 布什政府正准备寻求国会批准，将原本确定用于重建伊拉克被损毁的基础设施的33亿美元转向投入以建设法律和秩序为主的项目。

此句中 originally assigned... 这一部分为后置定语，修饰上文中的 $3.3 billion；同理，后面的 focused... 是用来修饰 programs 的。shift... into 表示"将……转向……"。

根据美国国会预算办公室的估计，伊战伊始，美国就为至少 500 万伊拉克人民备妥了人道救援物资，以每个人 500 美元计算，即花去 25 亿美元。而要使伊拉克基础设施恢复运转，在战后的头三年，每年至少要耗费 100 亿美元，至于重建学校、医院、博物馆等其他机构，大约还需 1,000 多亿美元。总经费最后甚至会超过 1,500 亿美元。

2 **The move comes against a background of steadily worsening public security in the country as it approaches a crucial first round of elections set for January.** (Para. 2) 随着定于 1 月举行的关键性首轮大选的临近，该国公共安全状况不断恶化，此次变化就是在这样的背景之下产生的。

此处所说的不断恶化的公共安全状况指的是伊拉克战后一直没有得到全面扼制的哄抢、破坏等活动，也包括愈演愈烈的袭击美军行动，这些都一直困扰着美国，成为重建恢复的主要障碍。

3 **Those working on the changes said the proposed reallocation amounts to nearly one-fifth of the $18.4 billion Congress approved last November to rebuild Iraq.** (Para. 3) 参与提出改变方案的官员指出，此次提出的重新分配资金数额占国会去年 11 月批准用于重建伊拉克的 184 亿美元总额的近五分之一。

美国国会 2003 年秋天批准向伊拉克提供 184 亿美元重建资金，但由于进展缓慢，到 2004 年底只花掉 10 亿美元多一点。白宫预算办公室将使用重建资金进展缓慢归咎于"伊拉克安全形势的急剧恶化"。

4 **At another level, however, the move indicates the administration's assessment that substantial changes are necessary to improve the security situations.** (Para. 5) 然而，从另一个层面上来看，这一变化意味着政府估计要进行重大改革以改善安全状况。

这句话中关系代词 that 引导同位语从句，用来补充说明先行词 assessment 的主要内容。

5 **The revised spending plan calls for adding 45,000 new recruits to the national police force, raising 20 new battalions to the current 42-battalion Iraqi National Guard, and beefing up the force that protects Iraq's long borders in an attempt to stop the flow of illegal traffic.** (Para. 7) 这一修改过的开支方案要求国家警察部队新征 45,000 名成员，在伊拉克国民警卫队现有的 42 个营基础上再增加 20 个营，并为保护漫长的伊拉克边境的部队补充力量，以阻断非法越境活动。

这句话中有三个动名词短语作并列宾语，adding...，raising... 和 beefing up...。

伊拉克国民警卫队为经过美国训练的准军事警察队伍。

D. Further Work on the Text

Write down at least three more comprehension questions of your own. Work in pairs and ask each other these questions. If you can't answer any of these questions, ask your classmates or the teacher for help.

1. _____

2. _____

3. _____

E. Vocabulary Practice

I. **Find the word that does NOT belong to each group.**

1. A. Congress   B. Washington   C. U.S. Embassy   D. State Department

2. A. election   B. recognition   C. organization   D. revision

3. A. reallocation   B. recognition   C. reconstruction   D. review

4. A. band   B. recruit   C. agency   D. focus

5. A. sewage   B. electricity   C. water   D. economy

II. **Complete the following sentences with appropriate words in their correct form.**

1. **propose, proposed (*a.*), proposal**

   1) The officials working on the changes said the _____ project was the result of comprehensive discussion.

   2) He _____ that we should go for a walk.

   3) It took him the whole night to write that grant _____.

2. **prior, priority, prioritize**

   1) Successful project teams recognize that not all requirements are created equal and, therefore, they need to _____ them and act accordingly.

   2) I have a _____ engagement and so can't go with you.

   3) The highest _____ of the government has been given to the problem of heavy traffic.

3. **approve, approved (*a.*), approval**

   1) My parents don't _____ of me smoking cigarettes.

   2) The resolution was _____ 82 to 16 with 18 abstentions.

   3) Mr. Ken is the most suitable person for this post for he is a man of _____ talent.

   4) He showed his _____ by smiling.

4. **assess, assessment**

   1) What is your _____ of the situation?

   2) The monthly income of school teachers in this place was _____ at $200.

5. **construct, construction, constructor, constructive, reconstruction**

   1) The new airport is still under _____.

   2) There are plans to _____ a new bridge across the river.

   3) The working conditions for the _____ were so poor that I couldn't help sympathizing with them.

   4) The Town Hall was destroyed in a fire two years ago. Its _____ is scheduled to start next month.

   5) Many _____ suggestions were made at the meeting.

Word Study

urge

n.　[常作 an urge] 强烈的欲望，迫切的要求，冲动：I had an urge to see him. 我迫切希望见到他。

v.　1. 鼓励，激励：He urged her to study physics. 他鼓励她学习物理。

　　　2. 催促，力劝：When my old friend Brian urged me to have a cigarette, I declined. 当我的老朋友布赖恩怂恿我抽烟时，我拒绝了。

　　　urge... into doing/to do 催促，怂恿……做：She urged me into taking steps in the matter. 她催我处理此事。

represent

v.　1. 描述，表现：This painting represents a storm. 这幅画描绘的是一场暴风雨。

　　　2. 代表，代理：We elected a committee to represent us. 我们选出一个委员会来代表我们。

　　　3. 象征，体现：The bald eagle represents the United States. 秃鹰象征美国。

派　representation, representative

absorb

vt.　1. 吸收，承受；承担：We will not absorb these charges. 我们不能承担这些费用。

2. 理解，掌握：To follow and absorb a newspaper article is a little bit hard as it requires a high level of mental involvement. 读懂报纸上的一篇文章比较吃力，因为这需要高度的脑力活动。

3. 吸引……的注意力，使全神贯注：The writer was absorbed in his writing and forgot his lunch. 作家全神贯注地写作，把午餐都忘了。

be absorbed in 专心致志于……：He is absorbed in study. 他专心学习。

派　absorption

shift

v.　1. 变换，改变：The wind shifted to the north. 风转向北吹。

2. 移动，转移：Shall I shift the chairs? 我可以把椅子移动一下吗？

n.　1. 替换，转变：The shift of power held back the social development of that country. 政权的更迭阻碍了那个国家社会的发展。

2. 班，轮班：Peter is on the day shift and I am on the night shift. 彼得上白班，我上夜班。

3. 推诿：You should face the consequences of the shift of responsibility. 你应该承担推托责任的后果。

link

v.　连接，联系：The two towns are linked by a railway. 一条铁路将这两个城镇连接起来。

n.　1. 环，节：A lot of links fitted together form a chain. 许多链圈连在一起组成链条。

2. 联系，纽带：Researchers have detected a link between smoking and heart disease. 研究人员发现心脏病与吸烟有关。

link... to 使……与……联系起来：The new bridge will link the island to the mainland. 新桥将把该岛与大陆连接起来。

派　linkage

Writing Practice

I. Complete the sentences based on the prompts given in brackets, using words instead of numbers and symbols.

Example:

How much is the vase worth? It's worth more than (£1 million) <u>one million pounds</u>.

1. (Thousand) _____ of dollars have been spent on this event.
2. There were (dozen) _____ of complaints after the broadcast.
3. For a lot of people ($100) _____ is a lot of money.
4. The whole collection at the National Gallery is worth (billion) _____.
5. Who's going to wade through a novel of more than (1,000) _____ pages?
6. My monthly income after tax is _____ ($2,656).
7. The population of that country is approximately _____ (22,230,000).
8. Student movements were very popular in the _____ (1960s).
9. You are the _____ (932) visitor to this website.
10. We've tried it _____ (5) times. Must we try it a _____ (6) time?
11. Could you please turn to the _____ (14) page?
12. Shall we meet at _____ (8:15) tomorrow morning?
13. What is the time now? _____ (18: 35).
14. China's population reached 400 million in _____ (1800).
15. Three _____ (+5=8).
16. How much is _____ (39-23)?
17. _____ (3×4) is twelve.
18. _____ (15÷3) is five.

II. **Choose the answer that best completes each of the following sentences.**

1. Our company sent surveys to 900 people, _____ have replied.
   A. of whom only 400 of them   B. of whom only 400
   C. only 400 of those who   D. only 400 who
2. Michael Harris was born _____.
   A. in the year 1986, at 11 p.m. on April 27th
   B. on April 27th at 11 p.m. in the year 1986
   C. at 11 p.m. in the year 1986 on April 27th
   D. at 11 p.m. on April 27th in the year 1986
3. David helps his mother with the housework every Saturday for about _____.
   A. one and half hours   B. a half and an hour
   C. an hour and half   D. one and a half hours
4. This is a _____ building, which is about _____ high.
   A. six-storey; 38 meter   B. six-storeys; 38-meter
   C. six-storeyed; 38 meters   D. six-storeys; 38-meters
5. —What would you like, sir?
   — _____.
   A. Two cups of coffees   B. Two coffees
   C. Two cup of coffees   D. Two cup coffee

6. The airport is _____ from my hometown.

   A. two hour's ride   B. two hours' ride

   C. two hour ride   D. two hours ride

7. _____ people around the world have come to realize the importance of learning Chinese.

   A. Hundred and thousand   B. One hundred thousand of

   C. A large amount of   D. Hundreds of thousands of

8. She was so ill that she had to ask for _____.

   A. a sick leave of three days   B. three-days sick leaves

   C. a three day sick leave   D. three-day sick leaves

9. The city wall is _____ and _____.

   A. 12 meter high… 8 meter wide

   B. 12 meters height… 8 meters width

   C. 12 meters in height… 8 meters in width

   D. 12 meter height… 8 meter width

10. I bought this shirt on sale. The original price was 49.99. It was 40% off. I only paid _____.

   A. 29.99   B. 19.99

   C. 24.99   D. 9.99

III. Put the following addresses into English.

1. 江苏省南京市北京东路21号5栋402室　210089

2. 上海世纪大道88号金茂大厦17层309室　200121

IV. Rewrite the sentences according to the models.

Model A:

Original sentence: Those who work on the changes said...

   New sentence: Those working on the changes said... (Text B)

1. People who wish to join the team may sign up here.

   _____

2. There are a lot of students who apply for scholarships.

   _____

Model B:

Original sentence: ... to shift \$3.3 billion which had originally been assigned to rebuild Iraq's shattered infrastructure into programs that were focused mainly to establish law and order.

   New sentence: ... to shift \$3.3 billion originally assigned to rebuild Iraq's shattered infrastructure into programs focused mainly to establish law and order. (Text B)

3. They imported lots of cars which were produced in Shanghai.

   _____

4. It was the first super computer which was designed by our own engineers.

   _____

Model C:

Original sentence: People expect that the shift of priorities will be reviewed at a meeting this week in Washington of senior officials from several government agencies.

New sentence: The shift of priorities is expected to be reviewed at a meeting this week in Washington of senior officials from several government agencies. (Text B)

5. People expect that Mr. Brown will be back before tomorrow morning.

   _____

6. According to the plan, all the students will meet on the playground this afternoon.

   _____

Model D:

Original sentence: At this time of discussing about renewing our labor movement, we must discuss the most urgent issue facing America and its working families.

New sentence: At this time of discussing about renewing our labor movement, how can we not discuss the most urgent issue facing America and its working families? (Text B)

7. With so many competent people working for us, we can surely finish the task on time.

   _____

8. Under such unique conditions, I simply couldn't do anything else.

   _____

V. **Combine each set of the sentences into one, using the connective words or expressions provided.**

1. a. One hundred and ten of these organizations have banded together.
   b. They have formed U.S. Labor Against the War (USLAW).
   c. USLAW is a national organization.
   d. It is committed to ending the war.
   e. It is committed to returning the troops.
   f. It is committed to restoring funding to social programs and government services.
   g. It is committed to changing the direction of U.S. foreign policy.
   New sentence: _____ (to, and) (Text A)

2. a. Union members and their family members are being killed, wounded, and disabled in a war.

   b. This war has already killed almost 1,500 U.S. military personnel.

   c. It has wounded more than 10,500 others.

   New sentence: _____ (that, and) (Text A)

3. a. The NWA should demand an immediate end to the U.S. occupation of Iraq.

   b. The NWA should demand return of U.S. troops to their homes and families.

   c. The NWA should demand the reordering of national priorities toward peace and meeting the human needs of our people.

   New sentence: _____ (and) (Text A)

4. a. It is weakening the rule of international law.

   b. It is not reinforcing the rule of international law.

   New sentence: _____ (rather than) (Text A)

5. a. There are too many idle men with little hope for work in this country.

   b. In such a country, job creation is closely linked to improved security.

   New sentence: _____ (where) (Text B)

6. a. The revised spending plan calls for adding 45,000 new recruits to the national police force.

   b. It calls for raising 20 new battalions to the current 42-battalion Iraqi National Guard.

   c. It calls for beefing up the force.

   d. The force protects Iraq's long borders in an attempt to stop the flow of illegal traffic.

   New sentence: _____ (and, that) (Text B)

7. a. The shift of priorities was initially drawn up at the U.S. Embassy in Baghdad.

   b. It is expected to be reviewed at a meeting this week in Washington.

   c. It will be a meeting of senior officials from several government agencies.

   d. Then the shift of priorities will go to Capitol Hill for final approval.

   New sentence: _____ (of, before) (Text B)

8. a. There is a recognition.

   b. The realities (in Iraq) are much different than they were last fall.

   c. Congress originally approved the budget last fall.

   New sentence: _____ (that, when) (Text B)

VI. Translate the following sentences into English.

1. 这几幅油画代表这个艺术家的早期风格。(represent)
2. 这种形势要求我们联合起来，对付我们共同的敌人。(call for)
3. 这场战争正逐渐耗尽我们的资源，使得我们的世界更加动荡不安。(drain away, secure)
4. 新任市长主张教师有更高的工资。(advocate)
5. 他渴望当一名电影明星，但随后又改变了主意。(urge)
6. 他提议我们去散步。(propose)
7. 人们相信肺癌与吸烟有一定的关系。(be linked to)
8. 委员会大多数成员愿意接受他们的提议。(be receptive to, proposal)

VII. Translate the following paragraph into Chinese.

Following the terrorist attacks of September 11, 2001, on the World Trade Towers in New York, the United States began a military campaign now known as the War on Terror. This effort has been used to justify a war and ongoing military operations in Afghanistan. It has also been used as justification for an invasion of Iraq, although there was no proven connection between the events of September 11 and Saddam's regime (政权) in Iraq.

VIII. Practical English Writing

Directions: Everybody wants to live in peace. But different people have different ideas about how to achieve peace. Some people believe that peace should be achieved by peaceful means while others maintain that only military force can lead to peace. What is your idea about this issue? Now discuss the issue with your friends. You can take some notes using the following table.

What is some people's idea about how to achieve peace? Why do they have this idea?	
What are other people's ideas about how to achieve peace? Why do they have these ideas?	
What is your idea about the issue? Give reasons to support your idea.	

Now write your essay here.

How Can We Achieve Peace?

People all over the world want to live in a peaceful environment. But different people have different views on how peace can be achieved. Some people believe that

_____

_____

_____

_____

_____

_____

_____

_____

_____

_____

10
Unit

Text A

Before Reading

Discuss the following questions in class.

1. What kinds of natural disasters do you know of? Have you personally experienced any of them?
2. Many people believe more and more natural disasters nowadays are caused by man. What's your opinion?

Reading

Prepare for Disaster and Save Lives

1   To save millions of lives, African nations and other developing countries need to anticipate[1] and reduce the risk from natural disasters, the UN Development Programme (UNDP) said in a report launched[2] on Tuesday in Nairobi, Kenya.

2   The report, titled "Reducing Disaster Risk: A Challenge for Development", focuses on how development is related to disaster risk, both positively and negatively[3]. It was also launched on Monday in Quito, Ecuador, and in Washington, D.C. and on Tuesday in Geneva.

3   The UNDP Bureau[4] for Crisis Prevention and Recovery[5] prepared the report that analyses global data from the past two decades, which indicates that death and destruction[6] caused by natural disasters in poor countries could be avoided by better planning and systematic risk analysis.

4   In a presentation on the report, the UNDP regional disaster reduction[7] adviser for Africa, Kenneth Westgate, said the report sought[8] to identify policies that could lead[9] to a reduction in the impact of natural disasters on the population.

5   "Disaster risk can be managed and reduced," he said. "The UNDP is suggesting that there is a real opportunity now to address disaster risk in a positive way to support poverty[10] reduction, the Millennium[11] Development Goals and all those other major development initiatives[12] that lay open to destruction by disaster events."

6   Describing the report as groundbreaking[13], the director of the UNDP's Dry Lands Development Centre, Philip Dobie, said for many years, people had dealt with the aftermath[14] of disasters but failed to recognise[15] that if properly managed, development could prevent natural events from becoming humanitarian[16] disasters.

7　　"Take, for instance[17], the case of floods. Over 196 million people in more than 90 countries are exposed[18] annually to the risk of floods, " he said. "However, floods turn into[19] humanitarian disasters mostly among poor communities[20], with low population densities[21], where disaster preparation and early warning are nonexistent[22]."

8　　Dobie said the natural disaster that affects Africa more than any other is drought[23]. "And drought is the most destructive disaster—one that creeps[24] up on us quietly and brings destruction to large sectors of society. "

9　　He added: "Fortunately, the message of today's report is that we can do a great deal to improve the situation. It is time to stop treating a disaster as an unexpected occurrence[25] that inevitably[26] causes death and destruction. It is time to recognise the role of development in reducing effects of disasters. "

10　　UN experts have described the report as the most extensive study ever published of global trends in exposure and risk to natural disasters.

11　　The report features a Disaster Risk Index[27] (DRI), which introduces a new method to measure global disasters and to enable experts to measure and compare physical exposure to risk between countries.

12　　The DRI demonstrates the link between human development and death rates following natural disasters. The most common natural disasters are tropical[28] cyclones[29], floods, earthquakes and drought.

New Words and Expressions

1　anticipate /æn'tɪsɪpeɪt/ v. 预先为……做准备

2　launch /lɔ:ntʃ/ v. 开始

3　negatively# /'negətɪvlɪ/ ad. 消极地，否定地

4　bureau★ /'bjʊərəʊ/ n. 局，办事处

5　recovery▲ /rɪ'kʌvərɪ/ n. 恢复，痊愈

6　destruction /dɪ'strʌkʃən/ n. 破坏，毁灭
　　destructive# /dɪ'strʌktɪv/ a. 破坏（性）的，毁灭（性）的

7　reduction▲ /rɪ'dʌkʃən/ n. 减少，降低

8　sought /sɔ:t/ seek 的过去式
　　seek to do sth. 试图做某事

9　lead to 导致

10　poverty /'pɒvətɪ/ n. 贫穷，贫困

11 millennium■ /mɪ'lenɪəm/ *n.* 千年，千禧年

12 initiative /ɪ'nɪʃɪətɪv/ *n.* 措施

13 groundbreaking# /'graund,breɪkɪŋ/ *a.* 突破性的，领先的

14 aftermath■ /'ɑ:ftəmæθ/ *n.* 后果，余波

15 recognise★ /'rekəgnaɪz/ *vt.* 明白，认识到

16 humanitarian■ /hju:,mænɪ'teərɪən/ *a.* 人道主义的

17 for instance 例如

18 expose /ɪk'spəuz/ *v.* 使处于……影响之下

19 turn into 变成

20 community /kə'mju:nətɪ/ *n.* 社区

21 density★ /'densətɪ/ *n.* 密集，稠密

22 nonexistent /nʌnɪg'zɪstənt/ *a.* 不存在的

23 drought■ /draut/ *n.* 干旱，旱灾

24 creep▲ /kri:p/ *vt.* 偷偷前进，缓慢地行进

25 occurrence▲ /ə'kʌrəns/ *n.* 发生的事情，事件

26 inevitably /ɪn'evɪtəblɪ/ *ad.* 不可避免的

27 index★ /'ɪndeks/ *n.* 指数，指标

28 tropical▲ /'trɒpɪkəl/ *a.* 热带的

29 cyclone■ /'saɪkləun/ *n.* 旋风，龙卷风

Proper Nouns

Nairobi /naɪ'rəubɪ/ 内罗毕 [肯尼亚首都]

Kenya /'kenjə/ 肯尼亚 [东非国家]

Quito /'ki:təu/ 基多 [厄瓜多尔首都]

Ecuador /'ekwədɔ:(r)/ 厄瓜多尔 [南美国家]

Washington, D.C. 华盛顿（市）[美国首都]（即哥伦比亚特区 [District of Columbia]）

Geneva /dʒɪ'ni:və/ 日内瓦 [瑞士城市]

Kenneth /'kenɪθ/ 肯尼思 [人名]

Westgate /'westgeɪt/ 韦斯盖特 [人名]

Philip /'fɪlɪp/ 菲利普 [人名]

Dobie /'dəubɪ/ 多比 [人名]

UNDP 联合国发展计划署

After Reading

A. Main Idea

Complete the following diagram with the sentences or expressions given below.

1. disaster risk can be managed and reduced
2. The DRI demonstrates the link between human development and death rates following disasters.
3. To save millions of lives, African nations and other developing countries need to anticipate and reduce the risk from natural disasters.
4. Reducing Disaster Risk: A Challenge for Development
5. stopping treating a disaster as an unexpected occurrence
6. a Disaster Risk Index
7. recognizing the role of development in reducing effects of disasters
8. The report analyses global data from the past two decades.

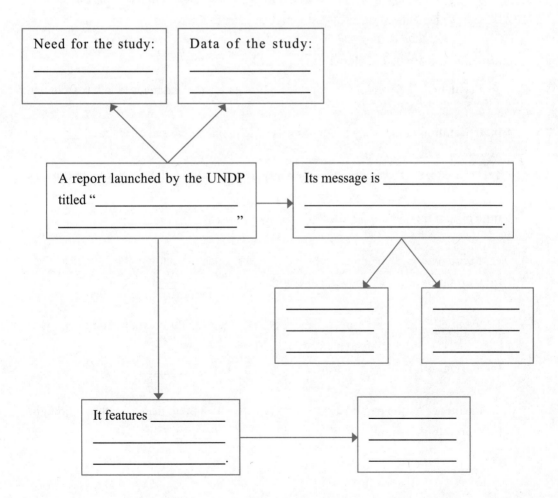

B. Detailed Understanding

I. Tell if the following statements are true (T) or false (F) according to the text.

1. _____ Death and destruction caused by disasters can sometimes be avoided.

2. _____ The UNDP report is negative about the situations of man in natural events.

3. _____ Floods turn into humanitarian disasters mostly among poor communities with low population densities.

4. _____ The most common natural disasters are tropical cyclones, floods, earthquakes and drought.

5. _____ With proper measures, developing countries can protect themselves from all natural disasters.

6. _____ Disaster Risk Index is a new method to measure global disasters.

7. _____ The UNDP report tries to ask for financial help from all over the world to reduce the impact of disasters.

8. _____ Hardly anybody treats a disaster as an unexpected occurrence.

9. _____ For many years, people had only dealt with the aftermath of disasters.

10. _____ Drought is the most destructive disaster in the world.

II. Explain the sentences by filling in the blanks.

1. **Text sentence:** ... developing countries need to anticipate and reduce the risk from natural disasters...

   Interpretation: ... developing countries need to prepare themselves against _____...

2. **Text sentence:** ... there is a real opportunity now to address disaster risk in a positive way to support poverty reduction...

   Interpretation: ... there is a real opportunity now to _____...

3. **Text sentence:** ... development initiatives that lay open to destruction by disaster events.

   Interpretation: ... development projects that _____.

4. **Text sentence:** ... if properly managed, development could prevent natural events from becoming humanitarian disasters.

   Interpretation: ... if development is properly managed, _____.

5. **Text sentence:** ... the natural disaster that affects Africa more than any other is drought.

   Interpretation: ... of all natural disasters, _____.

6. **Text sentence:** ... the message of today's report is that we can do a great deal to improve the situation.

   Interpretation: ... the main idea of today's report is that _____.

7. **Text sentence:** UN experts have described the report as the most extensive study ever published...

   Interpretation: According to UN experts, _____...

8. **Text sentence:** The report features a Disaster Risk Index...

   Interpretation: An important thing _____...

C. Detailed Study of the Text

1 **UN Development Programme (UNDP)** (Para. 1) 联合国发展计划署
   这是联合国的一个机构，在全球 166 个国家有分支机构，主要致力于建设民主政治、消除贫困、保护环境等。

2 **... focuses on how development is related to disaster risk, both positively and negatively.**
   (Para. 2) ······集中讨论发展和灾害风险之间正面与负面的联系。
   经济发展既可能减少也可能加重灾害风险，这取决于人类自身的观念与行为。

3 **... to support poverty reduction, the Millennium Development Goals and all those other major development initiatives...** (Para. 5) ······为支持减少贫困，新千年发展目标及所有其他主要发展计划······
   poverty reduction 及 the Millennium Development 皆为 UNDP 的工作项目，其中后者是该机构于 2002 年由联合国秘书长授权的一个新项目，旨在消除全球的贫困、饥饿和重大疾病。

4 **... the most extensive study ever published of global trends in exposure and risk to natural disasters.** (Para. 10) ······所有已公开发表的关于全球面临自然灾害风险趋势的研究中涉及面最广的一份。
   published 为过去分词，作 study 的后置定语。of 结构也修饰 study。

D. Talking About the Text

Work in pairs. Ask and answer the following questions first and then put your answers together to make an oral composition.

1. What did UNDP do recently?
2. What's UNDP's purpose of launching the report?
3. What does the report find from the data analysis?
4. How did people usually deal with disasters?
5. Which areas are most likely to be afflicted by floods?
6. Which natural disaster affects Africa the most and why is it the most destructive?
7. What is one of the features of the report?
8. According to Disaster Risk Index (DRI), what are the most common natural disasters?
9. How did UN experts describe the report?

E. Vocabulary Practice

I. Fill in the blanks with the new words or expressions from Text A.

1. More than 20 percent of the families in the country now live below the _____ line.
2. Now the whole world is under the threat of nuclear _____.
3. She made a quick _____ from the flu.
4. This new discovery is regarded as a _____ one.
5. Busy factories are an _____ of prosperity.
6. The program will _____ the creation of hundreds of new jobs.
7. She came home so thin and weak that her own children hardly _____ her.
8. Travel abroad _____ children to different languages and cultures.
9. John often _____ into the hall to listen to the singers.
10. Everyone is waiting for a _____ in the price of oil.

II. Complete the following dialogs with appropriate words or expressions from Text A.

1. A: The town is not exactly a cultural _____ like Paris in France.
   B: But we all think it's still attractive in many ways.
2. A: Jealousy is a very _____ emotion.
   B: And it's difficult for people to overcome.
3. A: Her success in the game seems totally _____.
   B: The truth is she has tried so hard for it.

4. A: The long period of _____ has caused a serious shortage of water supply in this area.

   B: Yes, but hopefully it will end next week.

5. A: You can't rely on her. _____, she arrived an hour late for an important meeting yesterday.

   B: Oh, that's too bad.

6. A: Tangshan City was badly hit in the 1976 _____.

   B: But today it has fully recovered.

7. A: I'm losing confidence in this job now.

   B: You've got to be more _____ about it.

8. A: The flood was the worst natural _____ to hit England for years.

   B: And it took away the lives of hundreds of people.

9. A: Disaster _____ is one of the main objectives of the meeting.

   B: It's usually more important than disaster recovery.

10. A: There is some _____ that fire would break out again.

    B: Then what should we do to avoid it?

 Before Reading

Discuss the following questions in class.

1. How can trade affect the environment?

2. Can development and environmental protection have positive rather than negative impact on each other?

 Reading

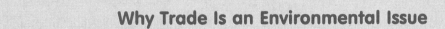

Why Trade Is an Environmental Issue

1   International trade is the mechanism[1] through which much of a country's environmental footprint[2] is imposed[3] beyond its own borders. The globalisation[4] of economic activities puts pressure on natural resources wherever environmental laws

are weak. The world's tropical rainforests[5], many wildlife species[6]—whose body parts are in demand for some folk medicines—and the once-clear streams and rivers flowing through New Zealand's export[7] dairying regions, are all examples of natural resources that are suffering greatly as a result[8].

2   The essential problem is the weakness of environmental laws, not trade itself. Nonetheless[9], trade can enlarge[10] the problem in two ways. First, trade opportunities create incentives for increased domestic production of goods that a country specialises[11] in. If production in a specialised economy is environmentally more harmful than a less specialised economy would be, and if environmental protection is missing[12] or inadequate[13], the gains from trade can be more than offset[14] by extra harm to the environment.

3   Second, trade opportunities lead business interests to lobby against effective environmental laws and policies, because they often see the additional costs cutting into[15] their competitive[16] advantage. This lobbying undermines[17] the capacity[18] of a community, or indeed a whole country, to move its economy on to an environmentally sustainable[19] basis.

4   Production subsidies[20] add element to the mix. In many countries, production is subsidized (overtly[21] or covertly[22]) by governments, which are responding to pressure from business and farming interests or seeking a trade advantage. Such subsidies distort[23] trade, and they usually have undesirable[24] environmental effects as well. That is particularly the case with the most widely-used, major subsidies—those for fossil[25] fuels, roading, agriculture[26], fisheries[27] and timber[28] production.

5   At a conceptual[29] level, two actions are needed to make trade environmentally sustainable. The first is to phase[30] out environmentally damaging subsidies that cause excessive production and distort trade. The second is to control the environmental impacts of production to a sustainable level, while ensuring that the cost of doing so is borne by the producer[31]. This will mean that the environmental costs of production are reflected in the price of the product (the polluter[32]-pays principle).

6   The polluter-pays principle also needs to be applied to global environmental problems such as climate change. This will require trade rules that allow for[33] the application of economically[34] sound environmental taxes and other measures on imports as well as domestic goods.

7   The future of trade liberalization[35], and the economic and environmental benefits it can bring, will depend on the ability of governments to establish trade rules and environmental policies that serve the common interest and not just the interests of business. At the same time, trade liberalization will have to help developing countries achieve sustainable growth, and not just serve the interests of developed countries.

New Words and Expressions

1 mechanism /'mekənɪzəm/ *n.* 机制

2 footprint# /'fʊtprɪnt/ *n.* 脚印，足迹

3 impose★ /ɪm'pəʊz/ *v. (on)* 把……强加于

4 globalisation# /ˌgləʊbəlaɪ'zeɪʃən/ *n.* 全球化

5 rainforest# /'reɪnˌfɒrɪst/ *n.* （热带）雨林

6 species▲ /'spiːʃiːz/ *n.* 种，类

7 export /ɪk'spɔːt/ *v.* 输出，出口

8 as a result 作为结果

9 nonetheless■ /ˌnʌnðə'les/ *ad.* 尽管如此，然而

10 enlarge /ɪn'lɑːdʒ/ *v.* 放大，扩大

11 specialise★ /'speʃəlaɪz/ *vi.* 专攻，专门从事

12 missing /'mɪsɪŋ/ *a.* 缺掉的，缺少的

13 inadequate# /ɪn'ædɪkwət/ *a.* 不足的，不充分的

14 offset■ /'ɒfset/ *vt.* 抵消

15 cut into 侵犯，损害

16 competitive★ /kəm'petətɪv/ *a.* （价格等）有竞争力的，竞争的

17 undermine■ /ˌʌndə'maɪn/ *vt.* 逐渐削弱，暗中破坏

18 capacity★ /kə'pæsətɪ/ *n.* 能力

19 sustainable# /sə'steɪnəbl/ *a.* 可持续的，持续性

20 subsidy■ /'sʌbsɪdɪ/ *n.* 补贴，补助金

    subsidize■ /'sʌbsɪdaɪz/ *vt.* 资助，给予津贴

21 overtly■ /'əʊvɜːtlɪ/ *ad.* 明显地，公开地

22 covertly■ /'kʌvətlɪ/ *ad.* 秘密地，悄悄地

23 distort /dɪs'tɔːt/ *v.* 使反常

24 undesirable# /ˌʌndɪ'zaɪərəbl/ *a.* 不受欢迎的，令人不快的

25 fossil■ /'fɒsəl/ *n.* 化石

26 agriculture★ /'ægrɪkʌltʃə(r)/ *n.* 农业，农学

27 fishery■ /'fɪʃərɪ/ *n.* 渔业

28 timber▲ /'tɪmbə(r)/ *n.* 木材，原木

29 conceptual# /kən'septjʊəl/ *a.* 概念的

30 phase★ /feɪz/ *vt.* 使……逐步进行

    phase out 逐步停止使用，逐步淘汰

31 producer# /prəʊ'djuːsə(r)/ *n.* 生产者，制造者

32 polluter# /pə'lju:tə(r)/ *n.* 制造污染者

33 allow for 考虑到，允许

34 economically# /ˌi:kə'nɒmɪkəlɪ/ *ad.* 在经济上

35 liberalization# /ˌlɪbərəlaɪ'zeɪʃən/ *n.* 自由化

Proper Noun

New Zealand /ˌnjuː'ziːlənd/ 新西兰 [大洋洲国家]

After Reading

A. Main Idea

Complete the following diagram with the expressions given below.

1. puts pressure on natural resources

2. establish trade rules and environmental policies that serve the common interest

3. the weakness of environmental laws

4. can be more than offset by extra harm to the environment

5. trade can enlarge the problem

6. distort trade and usually have undesirable environmental effects

7. may lead business interests to lobby against effective environmental laws

8. an environmental issue

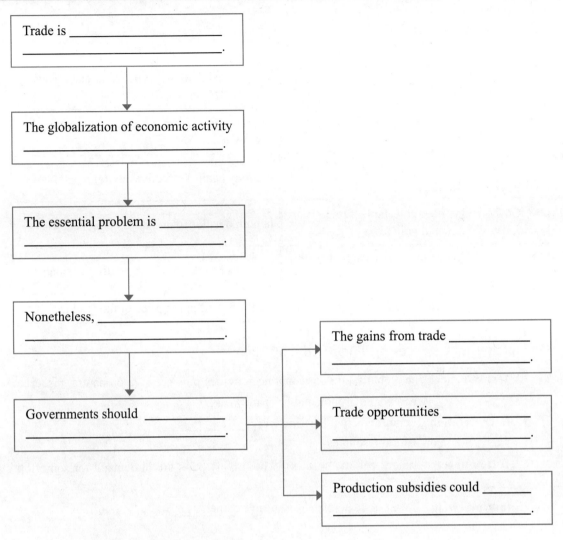

Trade is _____.

The globalization of economic activity _____.

The essential problem is _____.

Nonetheless, _____.

Governments should _____.

The gains from trade _____.

Trade opportunities _____.

Production subsidies could _____.

B. Detailed Understanding

I. **Make correct statements according to the text by combining appropriate sentence parts in Column A with those in Column B.**

Column A	Column B
1. The environmental costs of production should _____.	a. offset by extra harm to the environment
2. Trade liberalization will have to help _____.	b. puts pressure on natural resources
3. The globalization of economic activities _____.	c. cause excessive production and distort trade
4. The gains from trade can be more than _____.	d. that serve the common interest

(to be continued)

(continued)

Column A	Column B
5. Production subsidies usually _____.	e. increased domestic production that a country specializes in
6. Trade opportunities lead business interests to _____.	f. developing countries achieve sustainable growth
7. Environmentally damaging subsidies must be removed _____.	g. lobby against effective environmental laws
8. Governments need to establish trade rules and environmental policies _____.	h. both developed and developing countries
9. Trade opportunities create incentives for _____.	i. to make trade environmentally sustainable
10. Trade liberalization should help _____.	j. be reflected in the price of the product

II. **Explain the sentences by filling in the blanks.**

1. **Text sentence:** The globalisation of economic activities puts pressure on natural resources wherever environmental laws are weak.

   Interpretation: If environmental laws are weak, _____.

2. **Text sentence:** ... many wildlife species—whose body parts are in demand for some folk medicines...

   Interpretation: ... some folk medicines need _____ ...

3. **Text sentence:** ... the gains from trade can be more than offset by extra harm to the environment.

   Interpretation: ... there can be more harm to the environment than the gains from trade _____.

4. **Text sentence:** ... they often see the additional costs cutting into their competitive advantage.

   Interpretation: ... they often see they have less competitive advantage _____.

5. **Text sentence:** Production subsidies add an element to the mix.

   Interpretation: The situation gets more complicated _____.

6. **Text sentence:** That is particularly the case with the most widely-used, major subsidies...

   Interpretation: It is especially true _____...

7. **Text sentence:** The first is to phase out environmentally damaging subsidies...

   Interpretation: The first is to gradually _____ ...

8. **Text sentence:** The polluter-pays principle also needs to be applied to global environmental problems...

   Interpretation: To deal with global environmental problems, _____ ...

C. Detailed Study of the Text

----

1 **International trade is the mechanism through which much of a country's environmental footprint is imposed beyond its own borders.** (Para. 1) 通过国际贸易这一机制，一个国家对环境的影响大多延伸至了境外。

   这里是指一个国家在与别国进行贸易的过程中，常常会涉及当地的环境及自然资源，从而对该地区产生一定的影响。

   句中 footprint 字面意思是"足迹"，此处表示"影响"。

2 **... lead business interests to lobby against effective environmental laws and policies...** (Para. 3) ……使商业利益集团影响政府以阻止其制定有效的环境法规及政策……

   句中的 interests 意为"利益集团"、"业主"。下文第四段中的 interests 一词也是这个意思。lobby 原为名词，意为"门廊或公共建筑的接待厅"。在一些西方国家常有人集合在那里为某种立场或政策进行游说。这里 lobby 用作动词，意思是"游说"。

3 **... while ensuring that the cost of doing so is borne by the producer.** (Para. 5) ……同时要保证这样做的开支由生产方承担。

   so 这里指代前面的 to control the environmental impacts of production to a sustainable level。

4 **The future of trade liberalization, and the economic and environmental benefits it can bring, will depend on...** (Para.7) 贸易自由的未来，以及贸易自由所带来的经济与环境上的利益将取决于……

   句中的 it 指 trade liberalization。it can bring 为定语从句，修饰 the economic and environmental benefits。

D. Further Work on the Text

- -

Write down at least three more comprehension questions of your own. Work in pairs and ask each other these questions. If you can't answer any of these questions, ask your classmates or the teacher for help.

1. _____

2. _____

3. _____

E. Vocabulary Practice

- -

I. Find the word that does NOT belong to each group.

1. A. footprint   B. rainforest   C. offset   D. fossil

2. A. impose   B. missing   C. subsidize   D. enlarge

3. A. species   B. agriculture   C. fishery   D. roading

4. A. globalisation   B. protection   C. producer   D. liberalization

5. A. inadequate   B. nonetheless   C. overtly   D. greatly

II. Complete the following sentences with appropriate words in their correct form.

1. **export, exporter, exportation**

    1) Last year the company _____ more cases of wine to the UK.

    2) China is increasing its _____ to many developed countries.

    3) Japan is among the biggest _____ of electronic products in the world.

2. **adequate, inadequate**

    1) The government needs to take _____ measures to reduce disaster risks.

    2) The supply is _____ to meet the demand. More is needed for the completion of the project.

3. **lobby (*n.*), lobby (*v.*), lobbyist**

    1) The group is _____ for a reduction in defense spending.

    2) Wait for me in the hotel _____.

    3) When he came out of the meeting room, he was immediately greeted by a group of _____.

4. **subsidy, subsidize**

    1) Farming is greatly _____ by the government.

    2) When the workers retire from this factory, they can get _____ for health.

5. farm, farming, farmer

1) Joe had worked on the _____ all his life.

2) In some areas _____ lead a far better life than people in cities.

3) _____ is a business of growing crops or keeping animals.

Word Study

save

v. 1. 救，挽救：Thank you for saving my life. 感谢你救了我的命。

2. 储蓄，保留：So far, he has saved $5,000. 到现在他已经存了 5,000 美元了。

3. 节省，省去：We'll save a lot of time if we go by car. 如果我们乘车去可以省很多时间。

past

n. 过去，昔日，往事：There was something painful in his past. 他有痛苦的往事。

a. 以前的，过去的：From past experience she knew there was no need to ask him. 根据以往的经验，她知道没必要问他。

prep. 经过，超过：Will you go past my house on your way home? 你回去的路上会经过我家吗？

treat

v. 1. 对待，看待：She treats me like one of the family. 她把我像家人一样对待。

2. 处理：This problem must be treated immediately. 这件事必须立即处理。

3. 医疗，医治：Nowadays this disease can be treated with drugs. 现在这种病已经可以用药物治疗了。

4. 款待，请客：We treated him to lunch at the new restaurant. 我们在那家新开的饭店请他吃了午饭。

派 treatment

lead

v. 1.（为）带路：You lead and we'll follow. 你带路我们跟着来。

2. 领导，指引：An officer led his men into fight. 军官率领他的部下发起了战斗。

3. (*to*) 导致，通向：Does this road lead to the theatre? 这条路能到剧场吗?

4. 领先：Brazil led Germany 1-0. 巴西队 1 比 0 领先德国队。

n. 1. 领先，领导：The country has taken the lead in shipbuilding. 那个国家在造船业方面是领先的。

2. 铅：The lead pencil, which contains no lead, was invented in 1564. 不含铅的"铅"笔是 1564 年发明的。

派 leader, leadership

---

move

v. 1. 走动，移动：We'll have to move the bed closer to the wall. 我们得将床搬得离墙更近一些。

2. 感动，激动：I was deeply moved by their story. 他们的故事深深地打动了我。

3. 搬家，迁移：They've moved into a bigger office. 他们搬进了一间更大的办公室。

n. 动，动作，移动：He suddenly made a move towards the door. 他突然朝门口动了动。

move away 离开，搬家：His children moved away and he was left on his own. 他的子女都搬走了，只留下他一个人。

move on 继续移动，朝前走：That's enough rest—it's time to move on. 休息够了，继续前进。

move out 搬出去，（使）迁出：They want to move out of the old apartment. 他们想搬出旧公寓。

move up（使）晋升，向上：In this company, everyone wants to move up quickly. 在这家公司每个人都想迅速晋升。

派 movement, moving, remove

Writing Practice

I. **Choose the answer that best completes each of the following sentences.**

1. I wanted to borrow Jim's bike, but he refused _____.

   A. lending me one　　　　　　B. to lend it to me

   C. lend to me　　　　　　　　D. lend it to me

2. If Steve doesn't work harder at school he risks _____.

   A. failing the exams　　　　　B. fail the exams

   C. failing to the exams　　　　D. fail to the exams

3. Peter kept on asking me to go out with him, and eventually I agreed _____.
   A. meeting him after work   B. meet him after work
   C. to meet him after work   D. for meeting him after work

4. When I took my shoes back to the shop, they said that they couldn't give me a refund, but they offered _____.
   A. giving me a pair   B. to give me a pair
   C. to give me a new pair   D. to give me to a new pair

5. We were beginning to think we could never get out of the care, but finally we managed _____.
   A. to find a way to the exit   B. finding a way out
   C. find a way to the exit   D. finding its way to the exit

6. When Carrie was at school she was very keen _____ music and languages.
   A. for   B. at
   C. on   D. in

7. She was responsible _____ setting up the new standard for immigration.
   A. in   B. at
   C. for   D. with

8. Throughout his school years, Michael was very popular _____ his fellow students.
   A. of   B. with
   C. for   D. in

9. Gill knew that she was qualified _____ the job.
   A. in   B. at
   C. for   D. with

10. The interview panel were more concerned _____ Gill's ability to do administrative work.
    A. with   B. for
    C. in   D. at

11. _____ people trying to get into the football stadium.
    A. There were too much   B. There were too many
    C. It was too many   D. There was too much

12. I _____ you can swim so well and I can't.
    A. hate   B. hate it that
    C. hate that   D. hate it

13. Dave lost his job and was short of money, so _____ his apartment and move in with his brother.
    A. that he did was to sell   B. what he did was to sell
    C. what he did is selling   D. what he did were sell

14. _____ resigned, we would have been forced to fire him.
    A. Had he not   B. Had not he
    C. He had not   D. He had not had

15. _____ that Marie was able to retire at the age of 40.

    A. So successful her business was   B. So her business was successful

    C. So successful was her business   D. So was her successful business

16. _____ is hard to believe that Peter is already three years old.

    A. There   B. It

    C. That   D. This

17. There will be permanent settlements on the Moon one day. _____ is no dispute about that.

    A. This   B. That

    C. There   D. It

18. Having your own private swimming pool is an expensive luxury, although there is _____ that it is very pleasant to be able to swim whenever you want.

    A. no deny   B. no denying

    C. no denial   D. not denying

19. We really enjoyed visiting the Louvre when we were in Paris in spite of _____ to queue for two hours.

    A. having   B. have

    C. have had   D. had

20. In no way _____ be held responsible for this accident.

    A. Kevin can   B. Kevin could

    C. can Kevin   D. Kevin will

21. _____ had I tasted such wonderful coffee.

    A. No way   B. Only

    C. Never before   D. Barely

22. _____ was the ignorance of the disease at that time that sufferers were simply told to go to bed and rest.

    A. It   B. Such

    C. What   D. As

23. _____ change your mind, there will always be a job for you here.

    A. If   B. Had you

    C. Should you   D. Were you

24. _____ was Andy to move away from the city that he sold the house for much less than it was worth.

    A. So eager   B. Eager

    C. Eagerly   D. Eagerness

25. —We need new curtains. —Okay, let's buy _____.

    A. ones with flowers on   B. ones have flowers on

    C. some ones have flowers   D. some ones with flowers

26. —I don't suppose there'll be any seats left. —No, I _____.
    A. don't suppose   B. suppose
    C. don't suppose it   D. suppose not

27. They needed someone who was both an excellent administrator and manager. _____ was not easy to find.
    A. Such a person   B. A such person
    C. Such persons   D. Such person

28. —They could have been delayed by the snow. —Yes, they _____.
    A. could have   B. could do
    C. could be having   D. had been

29. The report is very critical and is clearly _____.
    A. intended to be so   B. intended to
    C. intended   D. intend to be

30. The party was excellent, and I'd like to thank all the _____.
    A. concerned people   B. concerning people
    C. people concerning   D. people concerned

31. My watch was among the _____.
    A. taken things   B. taking things
    C. things stolen   D. things stealing

32. We were delayed _____ an accident.
    A. because   B. because of there was
    C. because of   D. because it

33. _____, they slept soundly.
    A. Hot though was the night air   B. Hot as the night air was
    C. Hot although the night air   D. Hot although the night air was

34. If I _____ a more reliable car, I _____ to Chicago rather than fly.
    A. would have… would drive   B. had… had driven
    C. had… would drive   D. would have had… would drive

35. If he _____ a chance of success, he _____ to move to London.
    A. will have… would need   B. will have… will need
    C. were to have… will need   D. were to have… would need

36. I'd advise _____ chemistry class.
    A. to you for taking   B. you to take
    C. you taking   D. you took

37. You _____ insane if you think I'm going to lend you any more money.
    A. should be   B. are supposed to be
    C. must be   D. ought to be

38. I _____ happy to see him, but I didn't have time.
   A. will have been   B. would be
   C. will be   D. would have been

39. Jenny _____ leave the hospital only six hours after the baby was born.
   A. was able to   B. may
   C. can   D. is able to

40. The car broke down and we _____ a taxi.
   A. must have got   B. have to get
   C. had to get   D. must get

II. Rewrite the sentences according to the models.

Model A:

Original sentence: If development is managed properly, it could prevent natural events from becoming humanitarian disasters.

New sentence: If properly managed, development could prevent natural events from becoming humanitarian disasters. (Text A)

1. If you consider the problem seriously, it won't be too hard to solve.

   _____

2. If the young man had been given more care, he wouldn't have become a criminal.

   _____

Model B:

Original sentence: One feature of the report is its Disaster Risk Index.

New sentence: The report features a Disaster Risk Index. (Text A)

3. The murder of such a young child was a shock to the whole community.

   _____

4. I never took a risk of walking home alone at night.

   _____

Model C:

Original sentence: Some folk medicines need the body parts of many wildlife species.

New sentence: The body parts of many wildlife species are in demand for some folk medicines. (Text B)

5. The children greatly need her storybooks at the moment.

   _____

6. The only thing he needs now is time.

   _____

Model D:

Original sentence: The situation is particularly true of the major subsidies.

 New sentence: That is particularly the case with the major subsidies. (Text B)

7. This is not true of my friend.

 _____

8. Manufactured goods are especially like that.

 _____

IV. **Combine each set of the sentences into one, using the connective words or expressions provided.**

 1. a. The title of the report is "Reducing Disaster Risk."

  b. The report focuses on how development is related to disaster risk.

  New sentence:_____ (V+-ing) (Text A)

 2. a. The report sought to identify certain policies.

  b. Those policies could lead to a reduction in the impact of disaster.

  New sentence: _____ (that) (Text A)

 3. a. The director described the report as groundbreaking.

  b. The director said development could reduce disaster risk.

  c. Development is properly managed.

  New sentence: _____ (V+-ing, if) (Text A)

 4. a. Floods turn into humanitarian disasters mostly among poor communities.

  b. Some poor communities have low population densities.

  c. In some poor communities, disaster warnings are non-existent.

  New sentence: _____ (with, where)

   (Text A)

 5. a. International trade is a mechanism.

  b. Through international trade a country's environmental footprint is imposed beyond its own borders.

  New sentence: _____ (which) (Text B)

 6. a. The globalization of economic activities puts pressure on natural resources.

  b. In some places, environmental laws are weak.

  New sentence: _____ (wherever) (Text B)

7. a. Many wildlife species are suffering from international trade.

b. The body parts of these wildlife species are in demand for some folk medicines.

New sentence: _____ (whose) (Text B)

8. a. Trade opportunities lead business interests to lobby against effective environmental laws.

b. Business interests often see the additional costs cutting into their competitive advantage.

New sentence: _____ (because) (Text B)

V. Translate the following sentences into English.

1. 她患流感后很快就康复了。（recovery）
2. 这个地区的很多孩子都生活在贫困中。（poverty）
3. 她的来访完全出乎意料。（unexpected）
4. 目前的教学体系不会改变。（existent）
5. 这种虎已被宣布为珍稀物种。（species）
6. 他的进步表明了刻苦学习与好成绩的联系。（demonstrate, link）
7. 广告也是一项很具竞争性的产业。（competitive）
8. 父母不应该将自己的观点强加给孩子。（impose）

VI. Translate the following paragraph into Chinese.

In the northern hemisphere (半球), typically between January and March, snowstorms can bring life to a standstill. Drivers either cannot see through the fast-falling snow or their vehicles simply get stuck in deep snowdrifts (雪堆). Even when spring comes, things don't necessarily improve, as the melted snow can sometimes cause severe flooding.

VII. Practical English Writing

Directions: Every year, we hear about natural disasters that take the lives of hundreds and thousands of people and cause great damage to people's property. Now think of a kind of natural disaster that is common in your area or that you are familiar with. First explain what it is. Then describe the damage it can cause to human beings. Finally, think of some effective ways to reduce or eliminate its damage. You are encouraged to discuss with your classmates before you work on your essay. You can make use of the following table to take notes while discussing with your classmates.

What is the natural disaster you have thought of ?	
What is the damage the disaster causes?	
How to reduce or eliminate the damage?	

Now write your essay here. Remember to provide an appropriate title for your essay.

Glossary

A

abroad /ə'brɔːd/ *ad.* 在国外，到国外 U2

absentee# /ˌæbsən'tiː/ *n.* 不在者，缺席者 U4

absorb★ /əb'sɔːb/ *vt.* 吸收 U7

access /'ækses/ *n.* 享用权，享用机会 U1

accomplish★ /ə'kʌmplɪʃ/ *v.* 达到 U1
（目的），完成（任务）

accordingly▲ /ə'kɔːdɪŋlɪ/ *ad.* 照着， U6
相应地

accurately# /'ækjʊrətlɪ/ *ad.* 精确地； U2
准确地

acid▲ /'æsɪd/ *a.* 酸性的 U8

acquire★ /ə'kwaɪə(r)/ *vt.* 得到 U1

acquisition★ /ˌækwɪ'zɪʃən/ *n.* 获得，习得 U3

adaptive /ə'dæptɪv/ *a.* 适应的，有适 U4
应性的

addict★ /ə'dɪkt/ *vt.* 使上瘾，使入迷 U3

addicted /ə'dɪktɪd/ *a.* 上了瘾的，入了迷的 U3

adjust /ə'dʒʌst/ *v.* 调整，调节 U2

advancement /əd'vɑːnsmənt/ *n.* 发展， U3
进步

adventure /əd'ventʃə(r)/ *n.* 异乎寻常的经历 U4

advocate▲ /'ædvəkeɪt/ *vt.* 拥护，提倡， U9
主张

aerospace■ /'eərəʊspeɪs/ *a.* [只作定语] U5
航空与航天（空间）的

Afghanistan /æf'gænɪstæn/ 阿富汗 U8

aftermath■ /'ɑːftəmæθ/ *n.* 后果，余波 U10

agency★ /'eɪdʒənsɪ/ *n.*（政府等的） U9
专门行政部门

agenda★ /ə'dʒendə/ *n.* 议事日程 U1

aggressive▲ /ə'gresɪv/ *a.* 侵犯的，挑衅的 U2

agreement /ə'griːmənt/ *n.* 一致，感情融洽 U3

agriculture★ /'ægrɪkʌltʃə(r)/ *n.* 农业，农学 U10

aide■ /eɪd/ *n.* 助手，副官 U9

alienate■ /'eɪljəneɪt/ *vt.* 使疏远，离间 U3

alienation /ˌeljə'neɪʃən/ *n.* 疏远，离间 U3

allow for 考虑到，允许 U10

alphabet★ /'ælfəbɪt/ *n.*（一种语言的） U7
全部字母

alter /'ɔːltə(r)/ *v.* 改变 U3

alternative /ɔːl'tɜːnətɪv/ *n.* 供替代的选择 U4

anthrax■ /'ænθræks/ *n.* 炭疽（牛羊的 U6
传染病，常可致命，也可传染给人）

anticipate /æn'tɪsɪpeɪt/ *v.* 预先为……做 U10
准备

anticipate★ /æn'tɪsɪpeɪt/ *vt.* 预料 U6

anti-terrorism# /ˌæntɪ'terərɪzəm/ *n.* 反击 U6
恐怖主义

anxious /'æŋkʃəs/ *a.* 渴望的，急切的 U8

appellate■ /ə'pelət/ *a.* 上诉的 U6

appliance★ /ə'plaɪəns/ *n.* 工具，用具 U7

注： 无标记的单词表示B级词汇；★表示A级词汇；▲表示A级词汇之外的大学英语四
级词汇；■表示大学英语四级后词汇；#表示以常用词汇为词根所构成的派生词。

C

cap /kæp/ v. 胜过，超过 U5

capacity ★ /kə'pæsətɪ/ n. 能力 U10

Capitol Hill 美国国会山，（美国）国会 U9

carefree# /'keəfriː/ a. 无忧无虑的，轻松 U7
愉快的

CD-ROM abbr. [计]（信息容量极大 U7
的）光盘只读存储器

characteristic /ˌkærəktə'rɪstɪk/ n. 特性， U4
特征

chief /tʃiːf/ a. 主要的，首要的 U5

Chrysler /kraɪslə(r)/（美国）克莱斯勒 U6
汽车制造公司

circumstance ★ /'sɜːkəmstəns/ n. [pl.] U9
境遇，境况，经济状况

city-cell /'sɪtɪsel/ n. 市区 U5

civilian ▲ /sɪ'vɪljən/ n. 平民，百姓 U9

claim /kleɪm/ v. 声称，断言 U5

clarify ★ /'klærɪfaɪ/ vt. 澄清，阐明 U3

Cleveland /'kliːvlənd/ 克利夫兰 U2
[美国俄亥俄州东北部港口城市]

cliff▲ /klɪf/ n.（尤指海边的）悬崖， U8
峭壁

coach /kəʊtʃ/ vt. 训练，指导 U7

cognitive /'kɒgnɪtɪv/ a. 认知的 U3

Colin Powell /'kɒlɪn'paʊəl/ 科林·鲍 U9
威尔 [2001 年 1 月至 2005 年 1 月任美国
国务卿，身跨军政两界，历经四届美国
总统和无数次世界性危机]

commencement■ /kə'mensmənt/ n. 开始； U7
开端

commit to 对……作出承诺，承担义务 U9

community /kə'mjuːnətɪ/ n. 社区 U10

companionship /kəm'pænjənʃɪp/ n. 友谊， U3
交情

compatible /kəm'pætəbl/ a. 兼容的 U4

competitive ★ /kəm'petətɪv/ a.（价格 U10
等）有竞争力的，竞争的

completion# /kəm'pliːʃn/ n. 实现，完成 U9

complexity /kəm'pleksətɪ/ n. 复杂性 U3

compulsory■ /kəm'pʌlsərɪ/ a. 强制性的， U1
必须做的

conceivable▲ /kən'siːvəbl/ a. 可想到的， U6
可想象的

concept /'kɒnsept/ n. 概念 U6

conceptual# /kən'septjʊəl/ a. 概念的 U10

condemn▲ /kən'dem/ vt. 谴责 U9

conference ★ /'kɒnfərəns/ n.（正式）会议 U1

confirm /kən'fɜːm/ v. 进一步确定 U2

conflict /'kɒnflɪkt/ n. 冲突 U2

confront /kən'frʌnt/ v. 面临 U3

Congress ★ /'kɒngres/ n. 美国国会 U6

congressional# /kən'greʃənəl/ a. 国会的， U9
大会的

conscious /'kɒnʃəs/ a. 有意识的 U2

consequences /'kɒnsɪkwəns/ n. 后果 U7

conservation ★ /ˌkɒnsə'veɪʃən/ n.（对自然 U8
资源的）保护

conservationist# /ˌkɒnsə'veɪʃənɪst/ n. 自然 U8
环境保护主义者

consideration ★ /kənˌsɪdə'reɪʃən/ n. 考虑 U5

constant /'kɒnstənt/ n. 恒定的事物， U1
不变的事

construction project management 建筑工 U1
程管理

consult /kən'sʌlt/ v. 请教，查阅 U4

continual ★ /kən'tɪnjʊəl/ a. 不间断的， U3
不停的

contradiction* /ˌkɒntrə'dɪkʃən/ n. 矛盾　　U3

contribute /kən'trɪbjuːt/ v. 起促成作用　　U4

contribute* /kən'trɪbjuːt/ vi. (to) 是……　　U2
的部分原因

controllable# /kən'trəʊləbl/ a. 可控制的；　　U1
可支配的

convention▲ /kən'venʃən/ n. 公约，协议　　U8

convert* /kən'vɜːt/ vt. (使) 转变，(使)　　U7
转化

convey /kən'veɪ/ v. 表达 (感情等)　　U7

cooperation /kəʊˌɒpə'reɪʃən/ n. 合作，配合　U4

core /kɔː(r)/ a. 核心的，主要的　　U4

corporate■ /'kɔːpərət/ a. (法人) 团体　　U4
的，公司的

council* /'kaʊnsəl/ n. 委员会，理事会　　U1

covertly■ /'kʌvətlɪ/ ad. 秘密地，悄悄地　　U10

creation■ /kri'eɪʃən/ n. 创造，创建　　U9

creep▲ /kriːp/ vt. 偷偷前进，缓慢地行进U10

criminal /'krɪmɪnəl/ n. 罪犯，犯人　　U6

crisis /'kraɪsɪs/ n. 危机，危急关头　　U6

critical /'krɪtɪkəl/ a. 批评的，批判的　　U7

cross-cultural communication 跨文化交际　　U2

crucial* /'kruːʃəl/ a. 至关重要的，决定　　U9
性的

currently# /'kʌrəntlɪ/ ad. 现在，目前　　U8

cut into 侵犯，损害　　U10

cyberspace /'saɪbəˌspeɪs/ n. 网络空间　　U3

cyclone■ /'saɪkləʊn/ n. 旋风，龙卷风　　U10

D

dealer* /'diːlə(r)/ n. 商人　　U6

decrease /dɪ'kriːs/ v. 减小，减少　　U4

deficiency■ /dɪ'fɪʃənsɪ/ n. 缺点，缺陷　　U7

define /dɪ'faɪn/ v. 明确　　U7

definite /'defɪnɪt/ a. 明确的　　U1

deliver /dɪ'lɪvə(r)/ v. 提供 (服务)　　U1

democratic* /ˌdemə'krætɪk/ a. 民主的，　　U9
有民主精神 (或作风) 的

demonstrate* /'demənstreɪt/ v. 证明，说明U8

Dennis J. Reimer /'denɪs dʒeɪ'riːmə(r)/　　U6
丹尼斯·J. 赖默 [人名]

density* /'densətɪ/ n. 密集，稠密　　U10

deprive■ /dɪ'praɪv/ vt. (of) 剥夺，使丧失　U7

deserve* /dɪ'zɜːv/ v. 应受，应得　　U3

destination* /ˌdestɪ'neɪʃən/ n. 目的地，终点 U4

destruction /dɪ'strʌkʃən/ n. 破坏，毁灭　　U10

destructive# /dɪ'strʌktɪv/ a. 破坏 (性)　　U10
的，毁灭 (性) 的

digest* /dɪ'dʒest/ vt. 吸收，领悟　　U1

diploma■ /dɪ'pləʊmə/ n. 毕业文凭　　U1

disagreement# /ˌdɪsə'griːmənt/ n. 意见　　U3
不同，争执

disaster* /dɪ'zɑːstə(r)/ n. 灾难，大祸　　U8

discharge* /dɪs'tʃɑːdʒ/ n. 排放，放出　　U8

displace■ /dɪs'pleɪs/ vt. 取代，替代　　U7

disputant# /dɪs'pjuːtənt/ n. 争论的一方　　U2

dispute* /dɪs'pjuːt/ v. 争论，争吵　　U2

distinct* /dɪs'tɪŋkt/ a. 明确的，显著的　　U3

distort /dɪs'tɔːt/ v. 使反常　　U10

distort /dɪs'tɔːt/ vt. 扭曲，使变形　　U3

distribute /dɪ'strɪbjuːt/ v. 分配　　U7

district* /'dɪstrɪkt/ n. 区，地区，行政区　U9

Dobie /'dəʊbɪ/ 多比 [人名]　　U10

domestic* /dəʊ'mestɪk/ a. 家庭的，家用的 U5

downsize# /'daʊnsaɪz/ v. 裁员　　U4

draft▲ /drɑːft/ vt. 起草，草拟　　U8

drain away (使) 用尽　　U9

drain▲ /dreɪn/ vt. 使渐渐耗尽　　U9

drainage ■ /'dreɪnɪdʒ/ *n.* 排水，放水　　U8

drought ■ /draʊt/ *n.* 干旱，旱灾　　U10

E

eager /'iːgə(r)/ *a.* 热切的，渴望的　　U8

ecological# /ˌiːkə'lɒdʒɪkəl/ *a.* 生态的，生态学的　　U8

economically# /ˌiːkə'nɒmɪkəlɪ/ *ad.* 在经济上 U10

Ecuador /'ekwədɔː(r)/ 厄瓜多尔［南美国家］　　U10

effective /ɪ'fektɪv/ *a.* 有效的　　U2

effectively# /ɪ'fektɪvlɪ/ *ad.* 有效地　　U2

ego ■ /'iːgəʊ/ *n.* 自我，自己　　U4

electricity /ɪˌlek'trɪsətɪ/ *n.* 电　　U9

element /'elɪmənt/ *n.* 成分，要素　　U1

eliminate ★ /ɪ'lɪmɪneɪt/ *vt.* 根除，消除，排除　　U9

elsewhere /ˌels'hweə(r)/ *ad.* 在别处，到别处　　U4

embassy ★ /'embəsɪ/ *n.* 大使馆　　U9

emergency /ɪ'mɜːdʒənsɪ/ *a.* （仅用于名词前）应急的　　U8

emphasize /'emfəsaɪz/ *v.* 强调，着重　　U1

enact ■ /ɪ'nækt/ *vt.* 制定（法律）；通过（法案等）　　U8

encounter ▲ /ɪn'kaʊntə(r)/ *vt.* 遭遇，遇到　U7

end up 结束，告终　　U6

enhance /ɪn'hɑːns/ *v.* 提高，增强　　U4

enlarge /ɪn'lɑːdʒ/ *v.* 放大，扩大　　U10

ensure /ɪn'ʃʊə(r)/ *vt.* 确保，保证，担保　　U1

entry /'entrɪ/ *n.* 进入；入口处　　U2

environmental technology 环境技术　　U1

episode ▲ /'epɪsəʊd/ *n.* （连续剧的）一集 U5

Eric Rabkin /'erɪkˌræbkɪn/ 埃里克·拉　　U5

布金［人名］

erosion ▲ /ɪ'rəʊʒən/ *n.* 削弱，减少　　U9

establish /ɪ'stæblɪʃ/ *v.* 建立　　U2

evaluate ★ /ɪ'væljʊeɪt/ *v.* 评价，估价　　U6

evolution ★ /ˌiːvə'luːʃən/ *n.* 发展，演变　　U5

executive ★ /ɪg'zekjʊtɪv/ *n.* 主管，行政人员 U2

exhibit /ɪg'zɪbɪt/ *n.* 陈列品；展品　　U1

expertise /ˌekspɜː'tiːz/ *n.* 专门知识　　U2

exploit ★ /ɪk'splɔɪt/ *vt.* （为获取利益而）　　U8

利用

export /ɪk'spɔːt/ *v.* 输出，出口　　U10

expose /ɪk'spəʊz/ *v.* 使处于……影响之下 U10

exposure ★ /ɪk'spəʊʒə(r)/ *n.* 暴露　　U8

extended family （包括近亲的）大家庭 U3

extreme /ɪk'striːm/ *n.* 极端　　U4

eyewitness# /'aɪˌwɪtnɪs/ *n.* 目击者；见证人 U4

F

facelift ■ /'feɪslɪft/ *n.* （建筑物等的）　　U8

翻新，整修

facility /fə'sɪlətɪ/ *n.* [*pl.*] 设施，设备　　U8

fall into the hands of sb. 落到某人手里，成为某人所有　　U6

fascinating ■ /'fæsɪneɪtɪŋ/ *a.* 有极大吸引 U4

力的

federally# /'fedərəlɪ/ *ad.* 由联邦政府，通过联邦政府　　U6

federation ■ /ˌfedə'reɪʃən/ *n.* 联合会　　U9

fellowship ■ /'feləʊʃɪp/ *n.* 研究员基金　　U5

fetus ■ /'fiːtəs/ *n.* 胎儿，胚胎　　U7

figure out 想出，弄清……的原因　　U6

financial ★ /faɪ'nænʃl/ *a.* 财政的，金融的 U8

financially# /faɪ'nænʃəlɪ/ *ad.* （在）财政　　U4

方面，（在）金融方面

impact ▲ /'ɪmpækt/ n. 影响，作用　　　U6

impose ★ /ɪm'pəʊz/ v. (on) 把……强加于　U10

in essence 本质上，实质上　　　　　　U6

inability# /ˌɪnə'bɪlətɪ/ n. 无能力　　　U2

inadequate# /ɪn'ædɪkwət/ a. 不足的，不　U10
　　充分的

incentive■ /ɪn'sentɪv/ n. 刺激，鼓励　　U8

inclusion# /ɪn'kluːʒən/ n. （被）包括，包含 U8

index ★ /'ɪndeks/ n. 指数，指标　　　U10

induce ★ /ɪn'djuːs/ vt. 引起，导致　　U7

inevitably /ɪn'evɪtəblɪ/ ad. 不可避免的　U10

infinite ★ /'ɪnfɪnət/ a. 无限的，无穷的，　U3
　　不确定的

influential ▲ /ˌɪnflʊ'enʃəl/ a. 有影响的　U7

influx■ /'ɪnflʌks/ n. （人、资金或事　U8
　　物的）涌入，流入

infrastructure /'ɪnfrəˌstrʌktʃə(r)/ n. 基础设施　U9

inheritance# /ɪn'herɪtəns/ n. 从自然界承袭 U8
　　的共有资产（指水、土地、空气等）

initial ★ /ɪ'nɪʃəl/ a. 开始的，最初的　U4

initiate■ /ɪ'nɪʃɪət/ vt. 开始　　　　　U7

initiative /ɪ'nɪʃɪətɪv/ n. 措施　　　　U10

initiative /ɪ'nɪʃɪtɪv/ n. 主动的行动　　U8

injure /'ɪndʒə(r)/ v. （尤指在事故中）使　U5
　　（人、动物）受伤、弄伤

innovation /ˌɪnəʊ'veɪʃən/ n. 创新　　U4

innovative /'ɪnəʊveɪtɪv/ a. 富有革新精神的 U4

inspector# /ɪn'spektə(r)/ n. 检查员，监　U5
　　察员，巡视员

inspiration■ /ˌɪnspə'reɪʃən/ n. 给人以灵　U5
　　感的人（或事物）

install ★ /ɪn'stɔːl/ v. 安装　　　　　U2

intelligence ★ /ɪn'telɪdʒəns/ n. 智力，智　U1
　　慧，理解力

interact ★ /ˌɪntər'ækt/ vi. 互相作用，互相影响 U4

interim■ /'ɪntərɪm/ a. 临时的，暂时的　U9

interrupt /ˌɪntə'rʌpt/ v. 打断　　　　U3

intervention■ /ˌɪntə'venʃən/ n. 介入，干涉，U4
　　干预

Iraq /ɪ'rɑːk/ 伊拉克［西南亚国家］　U9

Iraqi /ɪ'rɑːkɪ/ a. 伊拉克的　　　　　U9

Irvine /'ɜːvɪn/ 欧文［地名］　　　　U5

isolate /'aɪsəleɪt/ v. 使孤立，使隔离　U9

J

Jane Nelsen /'dʒeɪn'nelsən/ 简·纳尔　U7
　　逊［人名］

jet /dʒet/ n. 喷气式飞机　　　　　　U6

John Sweeney /'swiːnɪ/ 约翰·斯威尼，　U9
　　美国最大工会组织美国劳工联盟及产
　　业工会联合会（劳联 - 产联）主席

joint ★ /dʒɔɪnt/ a. 共同的，共有的　　U3

Jonathan Baum /'dʒnəθən bɔːm/ 乔纳森·　U6
　　鲍姆［人名］

K

Kenneth /'kenɪθ/ 肯尼思［人名］　　U10

Kenya /'kenjə/ 肯尼亚［东非国家］　U10

Kingtron /'kɪŋtrən/ 金斯乔恩［机器人名］U5

knowingly# /'nəʊɪŋlɪ/ ad. 故意地，蓄意地 U6

L

lack /læk/ v. 缺乏　　　　　　　　U1

lag ★ /læg/ vi. 落后，走得慢　　　　U1

laser-rider# /'leɪzə'raɪdə(r)/ n. 激光骑行器 U5

launch /lɔːntʃ/ v. 开始　　　　　　U10

laundry /'lɔːndrɪ/ n. 待洗的衣服，洗好服 U5
　　的衣

novel /ˈnɒvəl/ a. 新的；新颖的　　　U1

nuclear ★ /ˈnjuːklɪə(r)/ a. 核子的，核能的 U5

numerous ★ /ˈnjuːmərəs/ a. 众多的，许多的 U7

nursery /ˈnɜːsərɪ/ n. 托儿所，保育室　　U7

O

occupation # /ˌɒkjʊˈpeɪʃən/ n. 占领，占据　U9

occupy /ˈɒkjʊpaɪ/ vt. 占，占据　　　　U5

occur /əˈkɜː(r)/ v. 发生，出现　　　　U4

occurrence ▲ /əˈkʌrəns/ n. 发生的事情，事件 U10

offence ★ /əˈfens/ n. 犯法行为，罪行　U6

offset ■ /ˈɒfset/ vt. 抵消　　　　　　U10

Okie /ˈəʊkɪ/ n.（美俚）俄克拉何马州人 U6

Oklahoma City 俄克拉何马城［美国俄　U6
　克拉何马州首府］

opposition # /ˌɒpəˈzɪʃən/ n. 反对，相反　U9

orient ★ /ˈɔːrɪənt/ vt. 使朝向，以……为方向 U4

outbreak # /ˈaʊtbreɪk/ n.（战事、情感、　U5
　火山等的）爆发

outcome /ˈaʊtkʌm/ n. 结果　　　　　U1

outgrowth # /ˈaʊtɡrəʊθ/ n. 产物，后果　U6

outstanding /ˌaʊtˈstændɪŋ/ a. 突出的，杰出的 U7

overseas /ˈəʊvəˈsiːz/ a. 海外的，国外的　U2

overtly ■ /ˈəʊvɜːtlɪ/ ad. 明显地，公开地 U10

owner /ˈəʊnə(r)/ n. 物主，所有人　　U4

Oxfordshire /ˈɒksfədʃɪə(r)/ 牛津郡［英国　U1
　英格兰郡名］

P

panel /ˈpænəl/ n. 陪审团，全体陪审员　U6

paperback # /ˈpeɪpəbæk/ n. 平装本，简装本 U5

payoff # /ˈpeɪɒf/ n. 回报，报偿　　　U8

peak ▲ /piːk/ n. 顶峰　　　　　　　U5

perceive /pəˈsiːv/ vt. 意识到，认识到　U7

perceive ★ /pəˈsiː v/ v. 感知，感觉　　U3

Ph.D.=Doctor of Philosophy 博士　　U1

phase out 逐步停止使用，逐步淘汰　U10

phase ★ /feɪz/ vt. 使……逐步进行　U10

Philip /fɪlɪp/ 菲利普［人名］　　　U10

pilot /ˈpaɪlət/ v. 驾驶（飞机）　　　U6

pint ▲ /paɪnt/ n. 品脱（液量单位）　U2

polluter # /pəˈljuːtə(r)/ n. 制造污染者　U10

positively /ˈpɒzətɪvlɪ/ ad. 积极地　　U7

postgraduate # /ˌpəʊstˈɡrædjʊət/ n. 研究生　U1

potential /pəʊˈtenʃəl/ a. 潜在的，有可能的 U2

pounding /ˈpaʊndɪŋ/ n. 猛击声　　　U8

poverty /ˈpɒvətɪ/ n. 贫穷，贫困　　U10

precaution ★ /prɪˈkɔːʃən/ n. 防备，警惕　U6

precede ■ /ˌpriːˈsiːd/ vt. 在……之前发生　U8
　（或出现）

predict /prɪˈdɪkt/ 预知，预言　　　U1

predictable /prɪˈdɪktəbl/ a. 可预言的；　U1
　可预报的

prevail ★ /prɪˈveɪl/ v. 流行，盛行　　U4

prevailing /prɪˈveɪlɪŋ/ a. 流行的，盛行的 U4

priceless # /ˈpraɪslɪs/ a. 珍贵的，无价的 U7

primary /ˈpraɪmərɪ/ a. 首要的，主要的　U2

priority /praɪˈɒrɪtɪ/ n. 优先权　　　U9

priority ★ /praɪˈɒrɪtɪ/ n. 优先考虑的事　U4

prodigy ■ /ˈprɒdɪdʒɪ/ n. 天才，奇才（尤　U7
　指神童）

producer # /prəʊˈdjuːsə(r)/ n. 生产者，制 U10
　造者

productivity ■ /ˌprɒdʌkˈtɪvətɪ/ n. 生产力，　U4
　生产率

profitability # /ˌprɒfɪtəˈbɪlətɪ/ n. 利润；利益 U4

progressive /prəʊˈɡresɪv/ a. 进步的，先进的 U4

S

Saturn /'sætən/ 萨图恩 [人名]　　　　U5

sb's good offices 某人的帮忙　　　　U2

scene /siːn/ n.（事故、案件等的）地　U5
　点，现场

science fiction ▲ n. 科幻小说　　　　U5

sci-fi /'saɪfaɪ/ n.（非正式）科幻小说　U5

seal ★ /siːl/ n. 印章　　　　　　　　U5

secretary /'sekrətərɪ/ n. 部长，大臣　U9

Secretary of State ［美国］国务卿　　U9

secure ▲ /sɪ'kjʊə(r)/ a. 安全的，无危险的　U9

seek to do sth. 试图做某事　　　　　U10

self-reliant /ˌselfrɪ'laɪənt/ a. 依靠自己　U7
　的，自力更生的

sensitive /'sensɪtɪv/ a. 敏感的，灵敏的　U2

sensitive /'sensɪtɪv/ a. 体贴入微的　　U7

sensitivity # /ˌsensɪ'tɪvətɪ/ n. 敏感性　U2

session ▲ /'seʃən/ n.（从事某项活动的）　U1
　一段时间

set sth. in motion 使某事物开始，导致　U6
　某事发生

settle ★ /'setl/ v. 解决　　　　　　　U3

sewage ■ /'sjuːɪdʒ/ n. 下水道（系统）　U9

sewer ■ /'sjʊə(r)/ n. 下水道　　　　　U9

shaky # /'ʃeɪkɪ/ a. 不可靠的，摇晃的，动　U9
　摇的

shatter ■ /'ʃætə(r)/ vt. 毁坏，使破灭　U9

shelf /ʃelf/ n. 架子，搁板　　　　　　U7

shipment ★ /'ʃɪpmənt/ n. 装载（或运输）　U6
　的货物

signer # /'saɪnə(r)/ n. 签名人，签字者　U9

significant /sɪg'nɪfɪkənt/ a. 影响深远的　U7

significant /sɪg'nɪfɪkənt/ a. 重大的　　U2

sincerity # /sɪn'serɪtɪ/ n. 真诚，诚挚　U7

situated # /'sɪtjʊeɪtɪd/ a.（尤指财政方　U8
　面的）处于……境地的

slave /sleɪv/ n. 奴隶，苦工　　　　　U5

Smith & Wesson 史密斯·韦森枪械公司　U6

sooner or later 迟早，早晚　　　　　U5

sort out 澄清　　　　　　　　　　　U3

sought /sɔːt/ seek 的过去式　　　　　U10

spacecraft ★ /'speɪskraːft/ n. 宇宙飞船　U5

spatial # /'speɪʃəl/ a. 空间的，与空间有关的　U7

specialise ★ /'speʃəlaɪz/ vi. 专攻，专门从事　U10

species ▲ /'spiːʃiːz/ n. 种，类　　　　U10

specific /spɪ'sɪfɪk/ a. 具体的　　　　U1

specifically ▲ /spɪ'sɪfɪkəlɪ/ ad. 特别地；　U9
　明确地

specify ▲ /'spesɪfaɪ/ vt. 明确说明，　U8
　具体指定

split ★ /splɪt/ vi. 分裂　　　　　　　U3

spokesrobot # /'spəʊksˌrəʊbɒt/ n. 机器人发　U5
　言人

squarely # /'skweəlɪ/ ad. 正好，不偏不倚地　U8

standstill # /'stændstɪl/ n. 静止状态，停止　U5

Starship Luna 星际飞船卢娜 [星际　U5
　飞船名]

starve ★ /staːv/ v. 使极度缺乏；（使）挨饿　U9

Statue of Liberty 自由女神像 [美国纽　U8
　约雕塑名]

statue ★ /'stætʃuː/ n. 塑像，雕像　　U5

status quo n. 现状　　　　　　　　　U4

steadily /'stedɪlɪ/ ad. 持续地　　　　U9

steadily # /'stedɪlɪ/ ad. 稳定地　　　U4

Steven Glenn /'stiːvənglen/ 斯蒂文·格　U7
　伦 [人名]

strain ▲ /streɪn/ n.（因用力过度而）　U5
　拉伤，扭伤

strategy ★ /'strætədʒɪ/ *n.* 策略　　　U2

strengthen ★ /'streŋθən/ *v.* 加强，巩固　U9

strike ★ /straɪk/ *vt.* 使（某人）突然意识到U3

subsidize ■ /'sʌbsɪdaɪz/ *vt.* 资助，给予津贴U10

subsidy ■ /'sʌbsɪdɪ/ *n.* 补贴，补助金　U10

substance /'sʌbstəns/ *n.* 真正的意义　U2

substantial ★ /səb'stænʃəl/ *a.* 可观的，大　U9
量的

sue ■ /sju:/ *v.* 控告，上诉　　　　　U6

suffer from 经受（疾病，不愉快之事等）U5

superb ▲ /sjʊ'pɜ:b/ *a.* 极好的，杰出的　U7

supplier # /sə'plaɪə(r)/ *n.* 供应者，供给者 U1

supportive # /sə'pɔ:tɪv/ *a.* 支持的；赞许的 U1

supposedly # /sə'pəʊzɪdlɪ/ *ad.* 根据推测　U7

survivor # /sə'vaɪvə(r)/ *n.* 幸存者，生还者 U6

sustainable # /sə'steɪnəbl/ *a.* 可持续的，　U10
持续性

T

takeover # /'teɪkˌəʊvə(r)/ *n.* 接管　　　U2

technical /'teknɪkəl/ *a.* 技术的，工艺的　U2

techno- /'teknəʊ/ 表示"技术"，"工艺"　U5
的前缀

tense /tens/ *a.* 紧张的　　　　　　　U2

tension ▲ /'tenʃən/ *n.* 紧张，紧张状态　U3

Terra-cotta Warriors 兵马俑　　　　　U8

the Ming Tombs 中国明代皇陵　　　　U8

Thomas Edison /'tɒməs'edɪsən/ 托马斯·　U7
爱迪生 [1847-1931，美国发明家]

threat ★ /θret/ *n.* 威胁，恐吓　　　　U6

threaten ★ /'θretən/ *vt.* 危及，对……构　U8
成威胁

thus /ðʌs/ *ad.* 因此，从而　　　　　U7

timber ▲ /'tɪmbə(r)/ *n.* 木材，原木　　U10

timing # /'taɪmɪŋ/ *n.* 时间选择；时机掌握　U2

Timothy J. McVeigh /'tɪməθɪ dʒeɪ ˌmək'veɪ/
蒂莫西·J. 麦克维 [人名]　　　U6

tone down （使）缓和；使协调　　　U2

tone ★ /təʊn/ *vi.* （尤指颜色）调和，和谐 U2

trace ★ /treɪs/ *vt.* 追溯，探索　　　　U5

tragedy ▲ /'trædʒɪdɪ/ *n.* 惨事，灾难　U6

treatment ▲ /'tri:tmənt/ *n.* 对待，待遇，处理U5

treaty ★ /'tri:tɪ/ *n.* （尤指国家间的）　U8
条约，协定

trivial ■ /'trɪvɪəl/ *a.* 琐碎的，不重要的　U5

troop ★ /tru:p/ *n.* [常 *pl.*] 军队，部队　U9

tropical ▲ /'trɒpɪkəl/ *a.* 热带的　　　U10

turn into 变成　　　　　　　　　　U10

typical /'tɪpɪkəl/ *a.* 典型的，有代表性的　U7

U

unanimously /ju:'nænɪməslɪ/ *ad.* 一致地，　U6
无异议地

unconventional # /ˌʌnkən'venʃənəl/ *a.* 非常　U6
规的

undergo ▲ /ˌʌndə'gəʊ/ *vt.* 经历，遭受　　U8

undermine ■ /ˌʌndə'maɪn/ *vt.* 逐渐削弱，　U10
暗中破坏

undesirable # /ˌʌndɪ'zaɪərəbl/ *a.* 不受欢迎　U10
的，令人不快的

UNDP 联合国发展计划署　　　　　U10

UNESCO /ju:'neskəʊ/ 联合国教育、科　U8
学及文化组织 [简称联合国教科文组织]

unfavourable # /ˌʌn'feɪvərəbl/ *a.* 不利的；　U1
不好的

unification # /ˌju:nɪfɪ'keɪʃən/ *n.* 统一，联合　U3

union ★ /'ju:njən/ *n.* 工会，联盟　　　U2

universal ★ /ˌju:nɪ'vɜ:səl/ *a.* 普遍的，共同的U8

urge /ɜːdʒ/ *vt.* 竭力主张，强烈要求　　U9

urgent /'ɜːdʒənt/ *a.* 紧迫的，紧急的　　U9

utopia■ /juː'təʊpɪə/ *n.* 乌托邦；理想的　　U3
　　完美境界

V

vaccine■ /'væksiːn/ *n.* 疫苗，菌苗　　U6

vending machine *n.* 自动售货机　　U2

veteran★ /'vetərən/ *n.* 老兵　　U6

vicious /'vɪʃəs/ *a.* 恶意的，恶毒的　　U6

virtual /'vɜːtʃʊəl/ *a.* [计] 虚拟的　　U3

visual★ /'vɪzjʊəl/ *a.* 视觉的，视力的　　U7

vital /'vaɪtəl/ *a.* 非常必需的，极其重要的 U7

vital /'vaɪtəl/ *a.* 重大的；紧要的　　U1

vitally /'vaɪtəlɪ/ *ad.* 重大地；紧要地　　U1

W

warfare■ /'wɔːfeə(r)/ *n.* 战争（状态）　　U6

Washington, D.C. 华盛顿（市）　　U10
　　[美国首都]（即哥伦比亚特区 [District of
　　Columbia]）

wear and tear 消磨，消耗，磨损　　U8

weatherproof# /'weðəpruːf/ *a.* 防风雨的，　　U8
　　不受气候影响的

weigh /weɪ/ *v.* 权衡　　U1

Westgate /'westgeɪt/ 韦斯盖特 [人名]　　U10

whiz /ʰwɪz/ *n.* 奇才　　U5

William Ury /'wɪljəm'jʊərɪ/ 威廉·尤　　U2
　　里 [人名]

willingly /'wɪlɪŋlɪ/ *ad.* 自愿地　　U6

withdraw /wɪð'drɔː/ *v.* 撤退　　U2

workplace# /'wɜːkpleɪs/ *n.* 工作场所　　U1

Y

yeah /jeə/ *ad.* （口）= yes　[变体]　　U4

yellowed /'jeləʊd/ *a.* 泛黄的，黄色的　　U5

Z

zone★ /zəʊn/ *n.* 地区，区域　　U8